Canine and Feline Dementia

Gary Landsberg • Aladár Maďari
Norbert Žilka
Editors

Canine and Feline Dementia

Molecular Basis, Diagnostics and Therapy

Springer

Editors
Gary Landsberg
North Toronto Veterinary Behaviour
Specialty Clinic
Thornhill
Ontario
Canada

Aladár Maďari
Univ of Veterinary Medicine and Pharmacy
Small Animal Clinic
Košice
Slovakia

Norbert Žilka
Slovak Academy of Sciences
Institute of Neuroimmunology
Bratislava
Slovakia

ISBN 978-3-319-53218-9 ISBN 978-3-319-53219-6 (eBook)
DOI 10.1007/978-3-319-53219-6

Library of Congress Control Number: 2017951451

Printed on acid-free paper

This Springer imprint is published by Springer Nature
The registered company is Springer International Publishing AG
The registered company address is: Gewerbestrasse 11, 6330 Cham, Switzerland

Preface

Canine Dementia: Unbearable Lightness of Ageing

The term "dementia" is generally used to describe a decline in mental ability of human beings which is severe enough to interfere with daily functioning. Human dementia represents a group of symptoms affecting memory, thinking and social abilities of patients with a decline in cognitive function beyond what might be expected from normal ageing. The probability of suffering from dementia progressively increases with age.

The question whether dogs can suffer from dementia similar to human patients has been raised by clinicians and basic researchers. Are they similar or are they representing a completely different pathological entity? From this perspective, the title of the book seems to be slightly provocative. In this text, authors from a wide field of research offer readers an in-depth insight into molecular biology, epidemiology, diagnostics and therapy of the disorder, commonly referred to in pets as cognitive dysfunction syndrome, and try to define the interface between the human and canine form of dementia. In fact, credit for much of the early studies that helped to shape our understanding of the field began with the collaboration of several research institutes studying the dog as a model for human brain ageing. The chapters by Joseph Araujo and Elizabeth Head provide us with insight into the similarities between canine brain ageing and human dementia. In both the dog and the human, the most prominent features of the disease are memory impairment, a decline in learning ability, alteration of social interactions and behavioural changes. The initial symptoms gradually worsen over time. As in human, age is the main risk factor.

Improvements in nutrition, preventive health screening and vaccinations, better hygiene and the development of antibiotics and other therapeutics for the successful treatment and prevention of disease have increased the life expectancy of dogs and cats. Thus, it is estimated that there are more than 45 million dogs around 7 years and older in the USA and Europe. However, as longevity increases, age-related health disorders related primarily to a decline in immune function and degenerative processes become increasingly more prevalent, including an increase in cognitively impaired dogs. Therefore, an increasing focus must be placed on management and therapy of age-related canine dementia. Identifying modifiable risk factors and developing preventive strategies for canine dementia represent important goals for veterinary medicine.

In terms of pathological hallmarks, the most prominent features are narrowing of gyri, widening of sulci and ventricular enlargement. The overall picture of the brain atrophy is strikingly similar to the early stages of human Alzheimer's disease. The characteristic microscopic changes are neuroinflammation, synaptic impairment and meningeal and parenchymal amyloid deposits. As in Alzheimer's disease (AD) patients, senile plaques are located mostly in the cerebral cortex. On the other hand, in contrast to human AD, senile plaques do not appear to progress beyond the diffuse plaque stage, and neurofibrillary lesions appear only rarely. The absence of classical hallmarks of AD pathology in canine and feline brains indicates that the underlying pathological processes might be different.

In the first few chapters, we describe the diverse clinical picture of canine and feline dementia, discuss the medical differentials for the disease and its phenotypic variability and introduce the dog as an appropriate animal model for human neurodegenerative disorders. We also shed light on the neuropathological hallmarks of canine and feline dementia. We highlight future perspectives of the modern diagnostic approaches based on the neuroproteomic technological progress. Finally, we address the current pharmacological and non-pharmacological approaches for therapy and discuss the risk and protective factors of feline and canine dementia.

It is the first comprehensive review of the topic of canine and feline dementia and brain ageing, which has become one of the major health problems in veterinary medicine in the last decades. It is intended for a wide readership including university teachers, students, scientists, clinicians and pet owners. We are pleased to have had the opportunity to edit and co-author this text with an international and diverse team of collaborators from both within and outside the field of veterinary medicine to share the most up-to-date knowledge of the field. It is our hope that this opus will attract further research across many fields of science to help unravel the mystery of canine and feline dementia and potentially further benefit our understanding of the human condition.

Thornhill, ON — Gary Landsberg
Bratislava, Slovakia — Norbert Žilka

Acknowledgements

As a veterinarian and specialist in veterinary behaviour, it is painfully clear that undesirable behaviour puts an emotional strain on the pet owner, both in having to manage the behaviour and to see the effects it has on their pet's welfare. In addition, many owners are not aware that a change in their pet's behaviour or the emergence of new behaviour problems is often the first or only sign of an underlying medical health issue. Yet, if reported and diagnosed promptly, it is often possible to improve or resolve the underlying problem or slow its progress. This is of particular importance in the senior pet, where behavioural changes are the clinical signs by which cognitive dysfunction is diagnosed. Therefore, it is essential for practitioners to include behaviour screening at each veterinary visit and to counsel owners of the critical importance of identifying and reporting signs.

Thank you for the love, support and understanding of my wife Susan, for not only this book but for what I have been able to accomplish in my veterinary career and in raising our three wonderful and successful children, Joanna, Mitchell and Jordan. Then there's Pepper, our little Havanese who brightens every day even when she challenges my skills as a behaviourist. I also recall with fondness, Buffy, Grace, Geil and Switch not only as companions but for all I have learned from them. In fact, in the book *Decoding Your Dog*, written for pet owners by diplomates of the American College of Veterinary Behaviorists, my chapter entitled "Growing Old with Grace" describes Grace, our Nova Scotia Duck Tolling Retriever whose cognitive dysfunction was effectively controlled from 12 to 16 years of age. Finally on a professional note, thanks to my colleagues Bill Milgram and Joseph Araujo, who have helped to guide me into and through the field of brain ageing and cognitive dysfunction, and to Norbert Zilka and Aladar Madari for their contributions to the field and without whose efforts and collaboration this book would not have been possible.

Thornhill, ON, Canada Gary Landsberg

I would like to gratefully acknowledge Prof. Michal Novak who, since we met 17 years ago, has always found the time to talk through my ideas, greatly benefiting my clarity of thought. I also wish to thank him for his inspiration and encouragement to pursue my career in the study of Alzheimer's disease and related neurodegenerative disorders.

I want to thank Aladar, Gary, Sagi and Joseph for all that they have taught me about canine cognitive dementia. It was their inspiring work that attracted me into the field. Special thanks go to Gary for his pioneering research without which this book would never have been written.

I am especially indebted to Petr Novak, M.D., Ph.D., who not only proof read our two chapters but also provided many stylistic suggestions and clarified our arguments.

My gratitude goes to the staff of the Institute of Neuroimmunology, who not only tolerated my attachment to the Institute but were also extremely welcoming to me from the outset; they have helped me in innumerable ways.

Finally, this book would not have been possible without the support of the exceptional Michaela Matisova, who has stood by me through good and bad times throughout the last 3 years.

Bratislava, Slovak Republic Norbert Žilka

I would like to mention the deceased Prof. Marán Kozak, who as a veterinarian, cynologist and award-worthy expert of canine soul and mind was my professional idol and showed me an incredibly beautiful area of behavioural and neurological veterinary medicine. I wish I could thank him as a pioneer of the behavioural field of veterinary medicine in Slovakia and for all the wisdom, experience and humanity with which he has supported and showed me the possibilities, ways and skills that over a period of 8 years since we met helped countless dogs and cats. His words and inspiration will always remain in my mind and heart.

My immense thanks go to a specialist in the research of Alzheimer's disease and associated neurodegenerative diseases Norbert Zilka, Ph.D., who had at any time provided my thoughts incredible understanding and whose friendly, professional and supportive approach has led to a deeper study of the problems of dementia in dogs and cats.

I am very grateful to Janka Farbakova, D.V.M, PhD. who is co-author of one of the chapters. She is not only excellent veterinary clinician but also a scientist with special interest in canine cognitive dysfunction. She has become an inseparable part of my professional and private life.

I want to thank Joseph, Gary and Sagi for sharing with me their experiences and knowledge in the field of behavioural medicine for dogs and cats.

Thanks also to the rector of the University of Veterinary Medicine and Pharmacy in Košice Prof. Janka Mojžišová for the strong support of my research.

Thanks to the team of the small animal clinic of the University of Veterinary Medicine and Pharmacy in Košice, people with whom we can work together on a daily basis in a very warm and friendly environment.

I want to thank my parents, brother and all my friends; without their love and support, I would not be able to reach any targets.

I must also mention my gratitude to my two dogs Matko and Fifinka and other animals that have helped me learn to understand them.

Košice, Slovakia Aladár Maďari

Contents

1 **Clinical Picture of Canine and Feline Cognitive Impairment** 1
Gary M. Landsberg and Rachel Malamed

2 **Behavioural and Medical Differentials of Cognitive Decline and Dementia in Dogs and Cats** 13
Sagi Denenberg, Francois-Xavier Liebel, and Jeremy Rose

3 **Phenotypic Variability and Clinical Staging of Canine Dementia** 59
Aladár Mad'ari, Petr Novak, and Norbert Žilka

4 **The Aged Dog as a Natural Model of Alzheimer's Disease Progression** 69
Joseph A. Araujo, Jessica Baulk, and Christina de Rivera

5 **Brain Aging in the Dog** 93
Jeff Nichol and Elizabeth Head

6 **Neuropathology of Feline Dementia** 103
Elizabeth Head and Danièlle Gunn-Moore

7 **Proteomic Approaches for Diagnostics of Canine and Feline Dementia** 113
Branislav Kovacech, Monika Zilkova, Jozef Hanes, and Rostislav Skrabana

8 **Current Pharmacological and Non-pharmacological Approaches for Therapy of Feline and Canine Dementia** 129
Sagi Denenberg and Gary Landsberg

9 **Preventive and Risk Factors of Canine Dementia** 145
Aladár Mad'ari, Jana Farbakova, and Norbert Žilka

Cognitive Dysfunction Syndrome: Information for Pet Owners 155

Clinical Picture of Canine and Feline Cognitive Impairment

1

Gary M. Landsberg and Rachel Malamed

Cognitive dysfunction syndrome (CDS) is a progressive, neurodegenerative disease of aged dogs and cats that manifests as behavioral changes, impaired learning and memory, awareness (response to stimuli), and confusion. CDS can severely impact animal welfare and the human-animal bond which may ultimately lead to shortened life span of the pet. Clinical signs in dogs and cats may include deficits in one or more categories. These categories represented by the acronym DISHAA include **d**isorientation, alterations in social **i**nteractions, changes in **s**leep-wake cycles, loss of **h**ousetraining and other learned behaviors, **a**ltered activity levels (increased or decreased), and increased **a**nxiety. Changes may also be seen with self-hygiene, appetite, and response to stimuli. CDS is diagnosed by exclusion of any medical and primary behavioral conditions, whose symptoms mimic that might be a cause of the signs. In addition, the presence of concurrent medical issues may confound a CDS diagnosis. Validated neuropsychological laboratory tests objectively quantify measures of learning and memory impairment that likely correspond to CDS signs. These tests provide a mechanism by which the effect of therapeutic agents can be assessed. However, as these tests require a trained personnel, standardized methodology, a cognitive assessment apparatus, and both time and consistency to assess the pet, they are a not a practical option in the clinical setting for family-owned pets. Although highly prevalent, signs of CDS are severely underreported to veterinarians by owners. Therefore, veterinarians must question owners proactively, obtain a thorough history, and screen for signs of CDS to ensure early identification which will then yield the greatest opportunity to treat early, slow decline, and improve CDS signs.

G.M. Landsberg (✉)
North Toronto Veterinary Behaviour Specialty Clinic, Thornhill, ON, Canada

CanCog Technologies, Toronto, ON, Canada
e-mail: gmlandvm@aol.com

R. Malamed
Behavior Consulting, Los Angeles, CA 90035, USA
e-mail: Behaviordogtor@gmail.com

G. Landsberg et al. (eds.), *Canine and Feline Dementia*,
DOI 10.1007/978-3-319-53219-6_1

1.1 Clinical Signs of Cognitive Dysfunction

Clinical signs of CDS have been described by the acronym DISH as a basis for screening and diagnosis. DISH refers to a **d**isorientation, altered social **i**nteractions with people or other pets, altered **s**leep-wake cycles, and **h**ouse soiling and loss of other learned behaviors (Azkona et al. 2009; Gonzalez-Martinez et al. 2011; Gunn-Moore et al. 2007; Madari et al. 2015; Landsberg et al. 2013; Landsberg et al. 2012; Osella et al. 2007; Rème et al. 2008; Neilson et al. 2001; Ruehl et al. 1995). Alternately, the use of the acronym DISHAA to include the categories of altered **a**ctivity levels and increasing **a**nxiety provides a more inclusive and potentially more sensitive screening questionnaire, based on studies that document alterations in these two clinical signs with increasing cognitive decline (Fast et al. 2013; Rosado et al. 2012). Other signs might include altered responsiveness to stimuli (i.e., exaggerated or reduced), altered interest in appetite or self-hygiene (i.e., increased or reduced), and altered control of feeding or drinking behaviors (Colle et al. 2000; Pugliese et al. 2005; Rofina et al. 2006) (Table 1.1).

Some dogs and cats might exhibit a single behavioral sign from one category, whereas others might exhibit multiple signs in a variety of categories. Similarly, there is a broad range in severity, including the number of presenting signs, number of affected categories, and intensity of signs that increases with advancing age (Azkona et al. 2009; Bain et al. 2001; Fast et al. 2013; Madari et al. 2015; Pugliese et al. 2005; Salvin et al. 2011a, b). In one study of dogs 8 years and older, signs that deteriorated most with age were activity and play levels, response to commands, and fears and phobias; however, for each of these signs underlying medical issues were a potential cause that had yet to be ruled out (Salvin et al. 2011a, b). In a study that examined the most common signs and their progression from non-affected to markedly affected, the most common signs in dogs with CDS were sleeping more during the day and restlessness at night (57%), altered social interactions (51%), disorientation (49%), and anxiety (46%). For dogs with mild cognitive dysfunction, the principal sign was increased daytime sleep (70%), with anxiety in 11%. In the non-CDS group, signs of anxiety were only reported in 4% of the dogs (Fast et al. 2013). In this study, the workup to rule out underlying medical causes was extensive including physical examination, blood, urine, and MRI where indicated. Madari and coworkers found that of the four categories of DISH, social interactions and sleep-wake cycles were most affected. In dogs with the most severe impairment, 67% were impaired in all four domains, while in the dogs with moderate dysfunction, 67% were affected in two domains (primarily social interactions and sleep-wake cycles), while with mild dysfunction, owners seldom reported changes unless asked with social changes being most affected (Madari et al. 2015). Azkona et al. reported alterations in social interactions and house training as the most commonly reported signs (Azkona et al. 2009). In another study, intermittent manifestations of anxiety were reported in 61% of dogs with CDS (Rème et al. 2008). In fact, because of their effects on the health and well-being of both the pet and the owner, signs of fear and anxiety may be among

Table 1.1 Senior canine and feline behavior questionnaire

SENIOR CANINE BEHAVIOUR SCREENING QUESTIONNAIRE

Date: ____________ **Owner:** ____________

Pet's Name: ____________ **Breed:** ____________

Weight: _______**lb / kg BCS (Out of 9)** ____ **Male** ___ **Neutered: Y**__ **N**__
Female ___ **Spayed: Y**__ **N**__

INSTRUCTIONS: The purpose of the questionnaire is to identify behavior changes or the onset of new behavior problems that have arisen since the age of 8 using the following key:
Scoring Key (severity): **0**=*none (no change)* **1**=*mild* **2**=*moderate* **3**=*severe*

BEHAVIOURAL SIGNS	**Score**
A. Disorientation	
- gets stuck, difficulty getting around objects, goes to hinge side of door	
- stares blankly at walls, floor or into space	
- does not recognize familiar people or pets	
- gets lost in home or yard	
- less reactive to visual (sights) or auditory (sounds) stimuli	
B. Social Interactions	
- more irritable / fearful / aggressive with visitors, family or other animals	
- decreased interest in approaching, greeting or affection / petting	
C. Sleep wake cycles	
- pacing / restless / sleeps less / waking at night	
- vocalization at night	
D. Housesoiling Learning and Memory	
- Less able to learn new tasks or respond to previously learned commands / name / work	
- Indoor soiling of urine __ or stools __ – decreased signaling to go out	
- Difficulty getting dog's attention / increased distraction / decreased focus	
E. Activity	
- Decrease in exploration or play with toys, family members, other pets	
- Increased activity – aimless pacing / wandering	
- Repetitive behaviors e.g. circling ___ chewing ___ licking __ star gazing	
F. Anxiety	
- Increased anxiety when separated from owners	
- More reactive / fearful to visual (sights) or auditory (sounds) stimuli	
- Increased fear of places / locations e.g. new environments / going outdoors	

Created by Dr. Gary Landsberg CanCog Technologies with support of Nestle Purina PetCare

the most common owner-presented behavior complaints in senior pets (Landsberg et al. 2010, 2011, 2013). These signs may include noise sensitivity, fear of people, fear of other animals, fear of new places or objects, hyperattachment, separation anxiety, and difficulty adapting to change. In some cases, anxiety may appear to be "generalized" with signs that exhibit in a variety of situations.

Another recent study further supports that senior dogs are more sensitive to emotional distress. Aged dogs (>7 years old) showed greater separation distress than did adult dogs (<7 years old) when separated from their primary caregiver. Although aged dogs showed a more passive response in a stranger separation test (SST), the physiological stress response as measured by salivary cortisol levels was higher in aged dogs (Mongillo et al. 2013b). Thus, fear and anxiety in senior pets may be associated with more passive (introverted) signs that are more likely to go underreported and undiagnosed if owners are not effectively educated in identifying and reporting these signs.

Dogs and cats with cognitive decline may also display the onset or progression of aggressive behavior toward household pets or family members. Pets with CDS may become the target of aggression by another household pet. In addition to an increase in anxiety and irritability, decreased awareness, perhaps in combination with sensory decline, may cause the pet to stumble upon another dog/cat or approach them when in possession of valued resources and locations. As well, pets with CDS may have concurrent health issues including sensory decline and pain and may not respond appropriately to social cues or warning signs. Senior dogs with CDS or other medical problems may require more attention or altered housing and care to address their infirmities, further contributing to instability in the dynamic among household pets.

In cats, the predominant sign in 11–14-year-old cats was altered social interactions, while changes in activity including aimless activity and vocalization were most common in cats 15 and over (Gunn-Moore et al. 2007, Landsberg et al. 2010, 2011).

1.2 Prevalence of Behavior Signs in Senior Pets

Prevalence of behavior signs in senior pets falls into two categories. Behavioral signs with the highest level of owner reporting but lower prevalence are those that have the greatest impact on the health, behavior, and well-being of the pet and its owner. Behavioral signs with the highest prevalence but lowest owner reporting are those that are sufficiently subtle or mild that they go unnoticed or are of minimal concern to the pet owner.

(a) Distribution of owner-reported behavior problems
In one study of 270 dogs over 7 years of age presented for behavior problems, 32% displayed aggression to family members, 16% aggression to family dogs, 9% barking, 8% separation anxiety, 6% disorientation and aggression toward unfamiliar people, 5% house soiling, 4% destructive and compulsive disorders, and 3% noise fears (Mariotti et al. 2009). Of 83 cats referred for behavioral consultations, most cats presented with marking or soiling (73%), followed by aggression (16%), vocalization (6%), and restlessness (6%) (Gunn-Moore et al. 2007). While CDS may be an underlying factor causing or contributing to many of these signs, other neurologic diseases, sensory impairment, endocrine and metabolic disorders, musculoskeletal disease, and other causes of pain must be ruled out.

(b) Prevalence of cognitive dysfunction syndrome
CDS is a highly prevalent yet grossly underreported condition of senior dogs and cats in which both prevalence and severity increase with age (Azkona et al. 2009; Bain et al. 2001; Gunn-Moore et al. 2007; Fast et al. 2013; Neilson et al. 2001; Madari et al. 2015, Salvin et al. 2010). While there is no cure, early detection and intervention may slow progression, prevent complications, and increase quality of life and life span.

Overall prevalence of CDS in dogs over 8 years of age has been reported to range from 14% to over 60% (Azkona et al. 2009; Neilson et al. 2001; Madari et al. 2015; Osella et al. 2007; Salvin et al. 2010). In a University of California, Davis, study of 180 owners of aged dogs, of those dogs 11–12 years of age, 28% were reported to have at least one category of DISHAA affected and 10% were positive for two or more categories, while in dogs 15–16 years old, 68% were positive for at least one category and 36% positive for two or more categories (Neilson et al. 2001). Twenty-two percent of dogs that did not have any signs developed signs 12–18 months later, while 48% of dogs with impairment in one category displayed impairment in two or more categories during this time period (Bain et al. 2001). Salvin and coworkers reported a prevalence of 5% in dogs 10–12 years of age, 23% aged 10–12, and 41% in dogs over 14 with an overall prevalence of 14.2% (Salvin et al. 2010). Katina et al. reported a prevalence of 13–16% for moderate to marked cognitive dysfunction in dogs aged 8–11 and 87%–100% in dogs > 13 (Katina et al. 2016). Madari et al. reported that of 300 dogs over 8 years of age, 159 of 215 dogs displayed signs of cognitive dysfunction after ruling out 85 dogs because of medical problems. Of these dogs, 42% with no impairment progressed to mild impairment, and 24% with mild impairment progressed to moderate over 6 months (Madari et al. 2015). After 1 year, 71.4% converted from none to mild impairment and 50% from moderate to severe (Madari et al. 2015). In yet another study in dogs over 8 years of age, over the course of approximately 3 years, 58% of those with no signs progressed to borderline CDS, and 11% of dogs moved from borderline to CDS (Fast et al. 2013). In a study of cats 11 years and older, 35% were diagnosed with CDS; this included 28% of 95 cats aged 11–15 years and 50% of 46 cats over 15 years of age (Gunn-Moore et al. 2007).

1.3 Pet Owner Reporting

As discussed, while many senior pets have signs of CDS, reporting is exceptionally low. In one study, 48% of pet owners reported at least one CDS sign in their senior dog 7 years of age or older, and only 17% of these owners informed their veterinarians (Proprietary market research 1999, Pfizer Animal Health). Another study found an estimated prevalence of 14.2% (68/479) in dogs greater than 8 years of age, but only 1.9% (9/479) of all dogs and 13% (9/68) of affected dogs had been diagnosed by a veterinarian previously. Madari et al. (2015) showed that pet owners did not usually report signs in dogs with mild cognitive impairment, until they were asked. Underreporting may be a result of owners who are unaware of the subtle signs or view them as untreatable, insignificant, or typical aspects of aging. Owners may not notify their veterinarians until signs advance to a point where they negatively impact the owner and it is evident that the pet is suffering. Therefore, to insure the earliest possible reporting of signs, veterinarians must be proactive in educating owners that any change in behavior or the emergence of new behavioral signs may be the earliest signs of medical conditions including pain, organ decline and dysfunction, endocrine disorders, and neurological conditions including CDS and that early diagnosis and treatment provide the greatest opportunity to improve signs, slow decline, and address pet welfare. In addition, to facilitate the earliest possible detection of CDS, veterinarians should consider the use of a broad-based screening questionnaire that includes all possible signs (Table 1.1) that can be provided to the client in advance of the visit or at the time of the visit.

1.4 Diagnosis

CDS, medical conditions and primary behavior problems must first be excluded (see Chap. 2). Neurological disease, sensory decline, musculoskeletal disease, and endocrine and metabolic disorders are the primary medical differentials (Gunn-Moore et al. 2007; Landsberg et al. 2010, 2011, 2012, 2013; Salvin et al. 2011a). While for most medical problems, concurrent medical signs or abnormal findings on physical and neurological examination and diagnostic tests are likely to be found, in some cases behavioral signs may be the first or only indication of an underlying health concern.

Since medical issues are common in the aged population and often coexist with CDS, this can be confounding. Medical issues, once diagnosed, may explain some behaviors, but concurrent CDS or other primary behavior problems may also contribute to a common sign. This necessitates continued investigation of primary behavior conditions other than those related to degenerative brain changes such as those that result from changes to the pet's environment and household composition. As dogs with CDS are more sensitive and less able to adapt to change, this may exacerbate or contribute to these issues.

Senior pets should be examined twice a year, for optimal screening of both medical and behavioral health. As discussed above, in dogs over 8 years of age, new behavioral signs may arise, and existing signs are likely to progress over the course of 6–12 months (Bain et al. 2001; Gonzalez-Martinez et al. 2014; Fast et al. 2013; Madari et al. 2015; Salvin et al. 2011b;). In a study of dogs over 8 years of age, changes in frequency and severity of behavioral signs were seen over the course of 6 months in over half of the behavioral signs, with the oldest dogs >12 years generally showing the greatest deterioration and highest change in frequency. More than two-thirds of responses showed significant age-related deterioration over 6 months with the greatest effects on play, activity, response to commands, and fears and phobias. As this was an Internet-based survey, further medical workup was required to determine if underlying medical pathology or cognitive dysfunction was responsible for the increased severity and frequency of the behavioral signs (Salvin et al. 2010).

Once signs of CDS are identified, the diagnostic workup should include a thorough medical history including the use of cognitive screening questionnaire, complete physical and neurologic exam, and laboratory/diagnostic tests. Video footage of the behaviors can help to describe, characterize, and determine frequency and time of occurrence. This is particularly important for the diagnosis and treatment of separation anxiety, attention-seeking behavior, compulsive disorders, and behaviors that occur with no obvious environmental trigger.

A complete blood count, serum biochemistry profile, thyroid level, and urinalysis should be obtained as a minimum database. Further diagnostic tests including endocrine testing, radiographs, ultrasound, and advanced imaging may need to be considered depending upon presenting signs and physical examination findings. While an MRI may demonstrate a decrease in total brain volume and hippocampal volume, frontal lobe atrophy, temporal lobe atrophy, ventricular enlargement, an increase in lesions in the frontal cortex and caudate nucleus, a decline in regional cerebral blood volume, and a decrease in diameter of interthalamic adhesions that might be consistent with a diagnosis of CDS, it is primarily indicated for ruling out other intracranial pathology that may mimic CDS (Hasegawa et al. 2005; Su et al. 2005; Tapp et al. 2004, 2006).

In order to determine if there is a primary behavior issue, the clinician should ask questions about the onset of the problem, frequency, time of the day, duration, and any other changes in health or behavior as well as assess a video recording of the problem. Information about the characteristics of vocalizations may help to determine motivation.

For animals who demonstrate night waking, attempts should be made to identify triggers or a triggering event, whether there has been an increase in reactivity in the pet's response to visual and audible stimuli and whether the pet has a chronic or more acute history of nocturnal signs. Evaluation of the pet's daily schedule including level of exercise and enrichment may determine if daytime activity is decreased or altered or if the pet is sleeping more during the day. A prior association with a fear-evoking event such as fireworks or thunder could cause the pet to display signs of anxiety at night in association with these events despite the absence of an obvious trigger. There may also be different diagnostic considerations for pets that have difficulty falling asleep, restless sleep, or waking early. Human responses will further influence the behavior (e.g., pet is rewarded with food, attention, toys, or outdoor access).

To differentiate between medical, behavioral, and environmental causes for house soiling, the history should include information about preferred locations, substrate, frequency, pattern (e.g., indiscriminate or specific location), litter box cleaning, and possible stressors within the home. Household changes or events that possibly correlate with the onset of increasing fear or anxiety should be identified. In addition to neurological disease and CDS, health issues that impact pain, irritability, altered mobility, and sensory decline (e.g., vision, hearing) might alter behavior or social signaling leading to displays of fear and anxiety.

1.5 Senior Cognition and Decline in Learning and Memory

While learning and memory deficits are likely to be among the earliest indicators of the pathology of brain aging and associated cognitive decline, the use of screening questionnaires is unlikely to be sufficiently sensitive for early identification by the average pet owner (Landsberg et al. 2011, 2013; Salvin et al. 2011b). In fact, other than those few behaviors that have been trained on cue, learning their name and housetraining, there are few trained behaviors on which most pets can be effectively assessed, and a moderate to marked decline might be expected before these behaviors might be affected. While a decline in learning and performance might be recognized at an earlier stage in dogs that have been trained for more complex tasks such as in working dogs, service dogs, or those trained for agility, the increased level of enrichment that their work provides may improve cognitive function and slow the onset of cognitive decline.

1.6 Neuropsychological and Behavioral Assessment

While pet owners may not begin to report behavior changes associated with cognitive decline until 11 years or older, deficits in learning and memory tasks in dogs and cats and functional changes in the neurons of the caudate nucleus in cats leading to impaired information processing have been demonstrated as early as 6 years of age (Araujo et al. 2005; Cotman and Head 2008; Levine et al. 1987a; Levine et al. 1987a, b; Pan et al. 2013; Salvin et al. 2010; Studzinski et al. 2006). These deficits have been shown to correlate with alterations in activity, social interactions and exploration, disorientation, house soiling and other learned behaviors, and sleep disturbances (Colle et al. 2000; Milgram et al. 1994; Rofina et al. 2006; Rosado et al. 2012; Siwak et al. 2001, 2003; Tapp et al. 2003). In addition, these neuropsychological models provide a standardized and validated means of assessing the efficacy of therapeutic drugs, diets, and supplements to slow decline or improve performance on these tasks (Araujo et al. 2012; Araujo et al. 2011; Cotman and Head 2008; Head et al. 2009; Milgram et al. 1994; Pan et al. 2010, 2013; Studzinski et al. 2006; Tapp et al. 2003, 2004).

Using a standardized test apparatus, several cognitive domains can be independently evaluated for age-related cognitive deficits (Adams et al. 2000a; Head et al. 1995; Milgram et al. 1994; Studzinksi et al. 2006; Zanghi et al. 2015). These tests involve training dogs and cats to use visual and/or spatial information to solve different problems. In simple discrimination learning, the animal is presented with two different objects, one of which covers a food reward. The rate of learning of simple discrimination tasks does not differ from younger dogs and cats (Adams et al. 2000a; b; Landsberg et al. 2009; Tapp et al. 2003, 2004). In the reversal task, reward contingencies are reversed such that food is now hidden under the object that was not previously rewarded (Tapp et al. 2003, 2004). In contrast to simple learning tasks, older dogs and cats require significantly more trials to learn to alter their response in this task compared to young dogs and are not readily able to modify learned behaviors, indicating dysfunction of the prefrontal cortex (Milgram et al. 1994; Landsberg et al. 2009; Tapp et al. 2003, 2004; Zanghi et al. 2015). Other behaviors that might be associated with prefrontal cortex dysfunction might include changes in personality including fearfulness and aggression, stereotypic pacing or circling, and a loss of previously learned behaviors, e.g., house soiling.

The aging process has also been demonstrated to affect spatial memory, which is measured by the ability of dogs to remember where they had last obtained a food reward. In fact, the level at which performance declines on memory tasks (e.g., DNMP) might be categorized to correspond to the stages of Alzheimer's disease (Adams et al. 2000a, b; Araujo et al. 2012; Head et al. 1995; Landsberg et al. 2009; McCune et al. 2008; Milgram et al. 1994; Pan et al. 2013; Studzinski et al. 2006; Zanghi et al. 2015). Clinically, this may present as wandering, getting lost, disorientation as well as disrupted sleep-wake cycles, and a decline in recognition of familiar people and animals.

Memory and spatial learning were also assessed in cats using a hole board task. Old cats (8–15) did not show differences in spatial learning compared to younger cats (<3 years old); however, memory errors were more abundant (McCune et al. 2008).

Adapting and validating these tasks for clinical use in pet dogs and cats are challenging as these tests are lengthy and complex and require trained personnel to administer (Gonzalez-Martinez et al. 2013; Heckler et al. 2014; Mongillo et al. 2013a; Nagasawa et al. 2012). However, one task, a food search test, may provide a methodology for pet owners to assess and track their pets in the home environment as it has been demonstrated to decline with age (>9 years) and with increasing cognitive dysfunction (but does not separate successful agers from those with cognitive dysfunction) (Gonzalez-Martinez et al. 2013).

Age-related behavioral differences have also been demonstrated in reactivity tests (Siwak et al. 2001, 2003; Rosado et al. 2012). The curiosity test assesses the dog's reaction and attention to objects (toys) in an open field arena. In this test, young dogs show significantly more exploration and contact with novel objects than old dogs, while cognitively impaired aged dogs showed the least object contact (Siwak et al. 2001). Cognitively, impaired dogs also demonstrated higher levels of locomotion than their age-matched unimpaired peers, which may be linked to stereotypy or wandering behavior (Siwak et al. 2001; Rosado et al. 2012). Assessment of exploratory behavior might offer a more practical measure of cognitive dysfunction in aged pets.

1.7 Summary

Clinical signs of CDS in dogs and cats as defined by DISHAA have a significant impact on the lives of pets and their owners. Laboratory tests are available to assess multiple domains that correlate to decreased capacity of learning and memory, spatial abilities, attention, psychomotor ability, and executive function. Many of these tests may not be practical for clinical use. However, client education and increased awareness to improve reporting rates as well as careful screening by veterinarians will aid in the early identification of CDS-like signs that may warrant further diagnostics to rule out physical causes that mimic CDS. Although there is no cure, early identification and intervention are critical to reducing the rate of cognitive decline and can mitigate the risk of premature end of life decisions while improving the quality and longevity of the pet's life.

References

Adams B, Chan A, Callahan H et al (2000a) Use of a delayed non-matching to position task to model age-dependent cognitive decline in the dog. Behav Brain Res 108(1):47–56

Adams B, Chan A, Callahan H, Milgram NW (2000b) The canine as a model of human cognitive aging: recent developments. Progr Neuro-Psychopharmacol Biol Psychiatry 24(5):675–692

Araujo JA, Studzinski CM, Head E, Cotman CW, Milgram NW (2005) Assessment of nutritional interventions for modification of age-associated cognitive decline using a canine model of human aging. Age 27(1):27–37

Araujo JA, Nobrega JN, Raymond R, Milgram NW (2011) Aged dogs demonstrate both increased sensitivity to scopolamine impairment and decreased muscarinic receptor density. Pharmacol Biochem Behav 98(2):203–209

Araujo JA, Faubert ML, Brooks ML, Landsberg GM, Lobprise H (2012) NOVIFIT®(NoviSAMe®) Tablets improve executive function in aged dogs and cats: implications for treatment of cognitive dysfunction syndrome. Int J Appl Res Veterinary Med 10(1):91–98

Azkona G, García-Belenguer S, Chacón G, Rosado B, León M, Palacio J (2009) Prevalence and risk factors of behavioural changes associated with age-related cognitive impairment in geriatric dogs. J Small Animal Practice 50(2):87–91

Bain MJ, Hart BL, Cliff KD, Ruehl WW (2001) Predicting behavioral changes associated with age-related cognitive impairment in dogs. J Am Veterinary Med Assoc 218(11): 1792–1795

Colle MA, Hauw JJ, Crespeau F, Uchihara T, Akiyama H, Checler F et al (2000) Vascular and parenchymal Aβ deposition in the aging dog: correlation with behavior. Neurobiol Aging 21(5):695–704

Cotman CW, Head E (2008) The canine (dog) model of human aging and disease: dietary, environmental and immunotherapy approaches. J Alzheimer's Dis 15(4):685–707

Fast R, Schütt T, Toft N, Møller A, Berendt M (2013) An observational study with long-term follow-up of canine cognitive dysfunction: clinical characteristics, survival, and risk factors. J Veterinary Intern Med 27(4):822–829

González-Martínez Á, Rosado B, Pesini P et al (2011) Plasma β-amyloid peptides in canine aging and cognitive dysfunction as a model of Alzheimer's disease. Exp Gerontol 46(7):590–596

González-Martínez Á, Rosado B, Pesini P et al (2013) Effect of age and severity of cognitive dysfunction on two simple tasks in pet dogs. Veterinary J 198(1):176–181

Gonzalez-Martinez A, Rosado B, Pesini P et al (2014) Cognitive dysfunction syndrome progress in dogs: a 6 month long study. In: ESVCE/ECAWBM Proceedings, Avignon, France, pp 97–98

Gunn-Moore D, Moffat K, Christie LA, Head E (2007) Cognitive dysfunction and the neurobiology of ageing in cats. J Small Animal Practice 48(10):546–553

Hasegawa D, Yayoshi N, Fujita Y, Fujita M, Orima H (2005) Measurement of interthalamic adhesion thickness as a criteria for brain atrophy in dogs with and without cognitive dysfunction (dementia). Veterinary Radiol Ultrasound 46(6):452–457

Head E, Mehta R, Hartley J et al (1995) Spatial learning and memory as a function of age in the dog. Behavioral neuroscience 109(5):851–858

Head E, Nukala VN, Fenoglio KA, Muggenburg BA, Cotman CW, Sullivan PG (2009) Effects of age, dietary, and behavioral enrichment on brain mitochondria in a canine model of human aging. Exp Neurol 220(1):171–176

Heckler MC, Tranquilim MV, Svicero DJ, Barbosa L, Amorim RM (2014) Clinical feasibility of cognitive testing in dogs (*Canis lupus familiaris*). J Veterinary Behavior: Clin Appl Res 9(1):6–12

Katina S, Farbakova J, Madari A, Novak M, Zilka N (2016) Risk factors for canine cognitive dysfunction syndrome in Slovakia. Acta Veterinaria Scandinavica 58(1):1

Landsberg G, Araujo JA, de Rivera C et al (2009) Validation of a cognitive test battery in cats. Abstract. In: Heath S (ed). In: Proceedings of seventh international veterinary behaviour meeting. ESVCE, Belgium, pp. 142–145

Landsberg GM, Denenberg S, Araujo JA (2010) Cognitive dysfunction in cats: a syndrome we used to dismiss as 'old age'. J Feline Med Surg 12(11):837–848

Landsberg GM, DePorter T, Araujo JA (2011) Clinical signs and management of anxiety, sleeplessness, and cognitive dysfunction in the senior pet. Veterinary Clin N Am: Small Animal Practice 41(3):565–590

Landsberg GM, Nichol J, Araujo JA (2012) Cognitive dysfunction syndrome: a disease of canine and feline brain aging. Veterinary Clin N Am: Small Animal Practice 42(4):749–768

Landsberg GM, Hunthausen W, Ackerman LA (2013) The effect of aging on behavior in senior pets. In: Behavior problems of the dog and cat, 3rd edn. Saunders, Edinburgh, pp 211–235

Levine MS, Lloyd RL, Hull CD, Fisher RS, Buchwald NA (1987a) Neurophysiological alterations in caudate neurons in aged cats. Brain Res 401(2):213–230

Levine MS, Lloyd RL, Fisher RS, Hull CD, Buchwald NA (1987b) Sensory, motor and cognitive alterations in aged cats. Neurobiol Aging 8(3):253–263

Madari A, Farbakova J, Katina S et al (2015) Assessment of severity and progression of canine cognitive dysfunction syndrome using the CAnine DEmentia Scale (CADES). Appl Animal Behav Sci 171:138–145

Mariotti VM, Landucci M, Lippi I et al (2009) Epidemiological study of behavioural disorders in elderly dogs. Abstract. In: Heath S (ed) Proceedings seventh international meeting of veterinary behaviour medicine, ESVCE, Belgium, pp 241–243

McCune S, Stevenson J, Fretwell L, Thompson A, Mills DS (2008) Ageing does not significantly affect performance in a spatial learning task in the domestic cat (*Felis silvestris* catus). Appl Animal Behav Sci 112(3):345–356

Milgram NW, Head E, Weiner E, Thomas E (1994) Cognitive functions and aging in the dog: acquisition of nonspatial visual tasks. Behav Neurosci 108(1):57–68

Mongillo P, Araujo JA, Pitteri E, Carnier P, Adamelli S, Regolin L, Marinelli L (2013a) Spatial reversal learning is impaired by age in pet dogs. Age 35(6):2273–2282

Mongillo P, Pitteri E, Carnier P, Gabai G, Adamelli S, Marinelli L (2013b) Does the attachment system towards owners change in aged dogs? Physiol Behav 120:64–69

Nagasawa M, Yatsuzuka A, Mogi K, Kikusui T (2012) A new behavioral test for detecting decline of age-related cognitive ability in dogs. J Veterinary Behav: Clin Appl Res 7(4):220–224

Neilson JC, Hart BL, Cliff KD, Ruehl WW (2001) Prevalence of behavioral changes associated with age-related cognitive impairment in dogs. J Am Veterinary Med Assoc 218(11): 1787–1791

Osella MC, Re G, Odore R, Girardi C, Badino P, Barbero R, Bergamasco L (2007) Canine cognitive dysfunction syndrome: prevalence, clinical signs and treatment with a neuroprotective nutraceutical. Appl Animal Behav Sci 105(4):297–310

Pan Y, Araujo JA, Burrows J, de Rivera C, Gore A, Bhatnagar S, Milgram NW (2013) Cognitive enhancement in middle-aged and old cats with dietary supplementation with a nutrient blend containing fish oil, B vitamins, antioxidants and arginine. Br J Nutr 110(01):40–49

Pan Y, Larson B, Araujo JA et al (2010) Dietary supplementation with medium-chain TAG has long-lasting cognition-enhancing effects in aged dogs. Br J Nutr 103:1746–1754

Pugliese M, Carrasco JL, Andrade C, Mas E, Mascort J, Mahy N (2005) Severe cognitive impairment correlates with higher cerebrospinal fluid levels of lactate and pyruvate in a canine model of senile dementia. Progr Neuro-Psychopharmacol Biol Psychiatry 29(4):603–610

Rème CA, Dramard V, Kern L, Hofmans J, Halsberghe C, Mombiela DV (2008) Effect of S-adenosylmethionine tablets on the reduction of age-related mental decline in dogs: a double-blinded, placebo-controlled trial. Veterinary Therapeut: Res Appl Veterinary Med 9(2):69–82

Rofina JE, Van Ederen AM, Toussaint MJM et al (2006) Cognitive disturbances in old dogs suffering from the canine counterpart of Alzheimer's disease. Brain Res 1069(1):216–226
Rosado B, González-Martínez Á, Pesini P et al (2012) Effect of age and severity of cognitive dysfunction on spontaneous activity in pet dogs—part 1: locomotor and exploratory behaviour. Veterinary J 194(2):189–195
Ruehl WW, Bruyette DS, DePaoli A, Cotman CW, Head E, Milgram NW, Cummings BJ (1995) Canine cognitive dysfunction as a model for human age-related cognitive decline, dementia and Alzheimer's disease: clinical presentation, cognitive testing, pathology and response to 1-deprenyl therapy. Progr Brain Res 106:217–225
Salvin HE, McGreevy PD, Sachdev PS, Valenzuela MJ (2010) Under diagnosis of canine cognitive dysfunction: a cross-sectional survey of older companion dogs. Veterinary J 184(3):277–281
Salvin HE, McGreevy PD, Sachdev PS, Valenzuela MJ (2011a) Growing old gracefully—behavioral changes associated with "successful aging" in the dog, *Canis familiaris*. J Veterinary Behav: Clin Appl Res 6(6):313–320
Salvin HE, McGreevy PD, Sachdev PS, Valenzuela MJ (2011b) The canine cognitive dysfunction rating scale (CCDR): a data-driven and ecologically relevant assessment tool. Veterinary J 188(3):331–336
Siwak CT, Tapp PD, Milgram NW (2001) Effect of age and level of cognitive function on spontaneous and exploratory behaviors in the beagle dog. Learn Memory 8(6):317–325
Siwak CT, Tapp PD, Zicker SC et al (2003) Locomotor activity rhythms in dogs vary with age and cognitive status. Behav Neurosci 117(4):813–824
Studzinski CM, Christie LA, Araujo JA, Burnham WM, Head E, Cotman CW, Milgram NW (2006) Visuospatial function in the beagle dog: an early marker of cognitive decline in a model of human aging and dementia. Neurobiol Learn Memory 86(2):197–204
Su MY, Tapp PD, Vu L et al (2005) A longitudinal study of brain morphometrics using serial magnetic resonance imaging analysis in a canine model of aging. Progr Neuro-Psychopharmacol Biol Psychiatry 29(3):389–397
Tapp PD, Siwak CT, Estrada J, Head E, Muggenburg BA, Cotman CW, Milgram NW (2003) Size and reversal learning in the beagle dog as a measure of executive function and inhibitory control in aging. Learn Memory 10(1):64–73
Tapp PD, Siwak CT, Gao FQ, Chiou JY, Black SE, Head E et al (2004) Frontal lobe volume, function, and β-amyloid pathology in a canine model of aging. J Neurosci 24(38):8205–8213
Tapp PD, Head K, Head E, Milgram NW, Muggenburg BA, Su MY (2006) Application of an automated voxel-based morphometry technique to assess regional gray and white matter brain atrophy in a canine model of aging. Neuroimage 29(1):234–244
Zanghi BM, Araujo J, Milgram NW (2015) Cognitive domains in the dog: independence of working memory from object learning, selective attention, and motor learning. Animal Cognit 18(3):789–800

Behavioural and Medical Differentials of Cognitive Decline and Dementia in Dogs and Cats

2

Sagi Denenberg, Francois-Xavier Liebel, and Jeremy Rose

Cognitive dysfunction syndrome (CDS) is a diagnosis of exclusion given that there is no specific diagnostic test or tool and that medical disorders can cause the same set of signs. The veterinary surgeon must first identify that signs are present, collect a full history and then perform a full physical examination and relevant diagnostic tests to rule out medical causes for the signs including blood and urine analysis, radiographs and diagnostic imaging such as magnetic resonance imaging (MRI) where indicated.

It is important to remember that with increasing age, there will be further degenerative changes leading to new behavioural signs that must be differentiated from CDS (Landsberg and Denenberg 2009). Only after these medical changes are detected, treated and controlled can the veterinary surgeon determine which of the signs might be caused by CDS. On the other hand, in the senior pet, medical problems and CDS can be present concurrently.

In this chapter we provide the veterinary surgeon with a list of possible differential diagnoses for CDS including behavioural, medical and neurological abnormalities. In addition, we aim to provide the reader with an explanation as to how they might confound or complicate the diagnosis.

2.1 Differential Diagnosis of CDS

2.1.1 Behavioural Differentials of CDS in Dogs and Cats

Senior pets may suffer from ageing-related decline such as reduced vision and hearing, medical abnormalities such as renal disease and pain (Landsberg and Denenberg 2009). All these can affect the pet's ability to cope with different environments and

S. Denenberg (✉) • F.-X. Liebel • J. Rose
Langford Vets, School of Veterinary Sciences, University of Bristol, Bristol, UK
e-mail: sagidvm@gmail.com; fl16868@bristol.ac.uk

G. Landsberg et al. (eds.), *Canine and Feline Dementia*,
DOI 10.1007/978-3-319-53219-6_2

situations. Reduced coping ability can later lead to behavioural changes such as the development or increase of fear, anxiety and aggression. Moreover, the pet may develop new strategies to cope. These can include avoidance, hiding and sleeping less to be better prepared.

2.1.1.1 Anxiety

Anxiety is the feeling of uneasiness or apprehension that a negative or dangerous situation may occur. Oftentimes anxiety is vague and not based on past experiences. Anxiety can occur in specific situations (e.g. owners are about to leave the house) or be more global where the pet is constantly anxious (Landsberg et al. 2001). Furthermore, anxiety can be mild where the pet may exhibit some anxiety-related behaviours or overwhelming where the pet outright refuses to interact, explore and learn. The behaviours and physiological signs of anxiety (Table 2.1) are modulated largely by serotonin and noradrenalin (both in the central nervous system (CNS) and peripherally).

Dogs and cats who are anxious may appear withdrawn to owners, leading to changes in interactions. Alternatively, increased anxiety may lead to more attention-seeking behaviours. Dogs may follow the owner around the home, show more separation distress and lie or rest only in areas where they can monitor the owner's presence. Cats may show attention-seeking behaviours as well. These may include following the owner, desire to sit on the owner's lap and even urine marking.

Pets may show more hypervigilant behaviours both at home and outside. This can include scanning the environment constantly, being easily aroused by minor triggers, disturbed sleep and high alertness (Ohl et al. 2008). The pet may eventually become exhausted and more irritable, leading to withdrawal from the environment and interactions.

At times these pets may eliminate inappropriately and lose interest in performing basic self-hygiene behaviours (mostly cats). The pet may be too anxious outside or near the litter box and begin to soil around the home or simply lose control over its sphincter at times of high anxiety (Landsberg et al. 2012). High and chronic anxiety can affect gastrointestinal (GI) motility, leading to diarrhoea. This in turn may lead to house soiling. Anxious cats may urine mark their environment trying to feel more secure.

Table 2.1 Behavioural and physiological aspects of anxiety

Behavioural changes	Physiological changes
Hypervigilance	Dilated pupils
High arousal	Tachycardia/tachypnoea
Attention-seeking behaviour	Excessive salivation
Monitoring of situation/owners	Panting
Changes in activity (increase of decrease)	Changes in GI motility leading to diarrhoea
Displacement behaviours	Muscle tremor/shaking
Behavioural inhibition (withdrawal)	Hypertension
Changes in sleep-wake cycles (light sleeping)	Increased Hypothalamic-pituitary-adrenal (HPA) axis activity

Anxious animals may have reduced appetite or avoid eating in different situations or environments, giving the impression of changes in appetite. For example, many dogs with separation distress will not eat from the time the owner is preparing to leave the house until the owner's return. They will, however, eat immediately upon the owner's return (Sherman and Mills 2008).

Chronic anxiety can lead to reduced production of several neurotransmitters that are required for effective learning and memory consolidation. Among these are serotonin, dopamine and brain-derived neurotrophic factor (BDNF). Serotonin and BDNF are particularly important for learning but also for inducing neurogenesis and synaptic plasticity (Gould 1999). Dopamine is important for memory consolidation in the hippocampus. Thus, pets with chronic anxiety may show deficits in memory or previously learnt behaviors during states of anxiety and will have difficulties acquiring new learning (Landsberg and Araujo 2005).

2.1.1.2 Fear

Fear is a basic survival response mechanism to a specific stimulus, an emotional response to a perceived threat and an innate response (Overall 2005). Fear is often adaptive, protecting the animal from harm. However, at times fear can become maladaptive; it may be overwhelming (panic attack), irrational (phobia) or out of context. Fear-related behaviours mostly originate in the amygdala and are modulated via the autonomic nerve system. The autonomic and somatic signs (Table 2.2) vary depending on the stimulus and its intensity; however, they can be grouped into defensive non-action (e.g. freezing, passiveness and hyper-attentiveness) or defensive action (e.g. flight or fight). The fear response is modulated primarily by noradrenalin both in the CNS and peripherally. In addition, the neurotransmitter gamma-aminobutyric acid (GABA) plays a role in fear-related behaviours in the amygdala.

Table 2.2 Behavioural and physiological signs of fear

Defensive action		Defensive non-action	
Behavioural	Physiological	Behavioural	Physiological
Vocalisation	Dilated pupils	Hiding	Dilated pupils
• Growling	Panting	Avoidance	Tachycardia or bradycardia later
• Barking	Tachycardia	Freezing	Changes in GI motility
• Whimpering	Tachypnoea	Muscle tremor	Pale mucous membranes
Cowering	Piloerection	Shaking	Hypertension
Lip curling	Hypertension	Submissive urination	HPA axis activation
Retracting lips	HPA axis activation	Rolling on side and exposing the abdomen	
Staring or gazing away		Whimpering	
Lunging or retreating		Closing eyes or gazing away	
Biting			

Ageing can lead to changes in fear responses in animals as a result of a decreased ability to cope with different triggers. For example, a dog with reduced hearing or vision and degenerative joint disease is not capable of avoiding a fearful situation as in the past and as a result may attack or freeze. This can easily be viewed as changes in interaction with the same trigger (Landsberg and Denenberg 2009).

Animals may be fearful of another pet in the house that controls access to resources such as food, sleeping areas and soiling areas. The fearful pet may eat less, soil inappropriately and even sleep less.

Sudden and intense fear may lead to loss of control over the bladder and anal sphincter. Dogs and cats that experience this may soil inappropriately. A dog who has fireworks phobia and in the past could have run away and hid may no longer be able to do that and may void its bladder as a result of intense fear (Sherman and Mills 2008).

Pets who learn that certain situations, environments or people cause fear may avoid the area or the interactions. These pets may also refuse to perform certain tasks when asked by the owner. This in turn can be viewed as forgetfulness.

2.1.1.3 Night-Time Waking

Dogs and cats tend to adopt their owners' routine and schedule. However, cats typically have dusk and dawn activity patterns (crepuscular species) related to their food acquisition needs and the activity of prey. Older pets may become more active at night-time when the owner is sleeping (Landsberg and Denenberg 2009). Younger pets may also wake their owners at night; however, the problem is more common in senior animals, because of many health issues (e.g. renal, gastrointestinal and joint diseases) that may contribute. They may seek food or attention or need to eliminate. In addition, the pet may wake up and walk around the house, and the owner's reaction (trying to stop or comfort the pet) may actually reinforce this behaviour by giving food or attention. Common reasons for night-time waking include not only CDS but also conditions that cause pain, discomfort, confusion or unsettled sleep including arthritis, gastrointestinal disease, organ-related dysfunction and endocrine disorders such as diabetes mellitus, hyperthyroidism and hypertension. Sensory decline (vision, hearing) may reduce the pet's awareness and ability to cope with darkness or its response to stimuli. Any disease process that might increase thirst or hunger or frequency or volume of urine or stool may also contribute to night waking. Changes in sleep-wake cycles are common CDS-presenting complaints (Landsberg et al. 2011, 2013).

A diagnosis is established by performing the physical examination, blood and urine tests, blood pressure measurements, pain assessment and thorough history collection and observing behaviour patterns. In addition, investigating other possible concurrent signs and problems must take place in order to rule out cognitive decline. These signs may include increased anxiety and fear; changes in appetite, elimination routine, and social interactions; and increased vocalisation (Denenberg and Landsberg 2016).

Changes in daily routine, including limiting cats from going outdoors or altering the dog's walk and play schedule or feeding routine, and household changes (e.g. adding new pets, loss of another companion pet or owners, moving and renovation) might lead to decreased sleep quality and duration. Inciting factors (e.g. disease, outside animals or routine changes) might be different from maintaining factors (e.g. owner's reinforcement). When the problem becomes chronic, it may be difficult to reverse the routine; therefore, prompt action is required (Landsberg et al. 2013).

Anxiety in senior pets (see above) may lead to development of displacement behaviours, one of which is repetitive pacing. While this can be throughout the day, it might only be noticeable to owners in the evening or nights when they are at home or trying to sleep (Landsberg et al., 2011; Landsberg and Denenberg 2009).

2.1.1.4 Excessive Vocalisation

Vocalisation becomes excessive or is considered to be a problem when it occurs at inappropriate times (e.g. nights, baby napping) or is particularly loud or long (Denenberg and Landsberg 2016). Duration, frequency and time of the day are some of the parameters that should be evaluated. Moreover, changes in health and other behaviours should be evaluated as well. Thorough history should be taken to evaluate any patterns or triggers (e.g. outdoor animals, being left alone, confined near feeding times or fearful situations).

Attention given by the owner might reinforce the behaviour. Punishment may reduce the behaviour for a short period of time; however, it is likely to increase anxiety and fear, leading to increased vocalisation in the long run, and does not address the underlying motivation to vocalise (Landsberg et al. 2011).

Painful conditions (e.g. arthritis, gastrointestinal, dental or neuropathic), organ dysfunction (e.g. chronic renal disease with uraemia or liver disease) and central nervous system disease including cognitive dysfunction, sensory decline and thirst or increased hunger might lead to increased vocalisation (Denenberg and Landsberg 2016). Anxiety can be coupled with increased vocalisation in the older pet, and this could be due to health issues, an increased dependency on the owner for comfort and safety, and sensitivity to changes in the environment.

2.1.1.5 House Soiling

Senior pets may soil inappropriately due to cognitive decline. However, there are many other reasons that should be considered first. The first question is whether the pet was housetrained in the past or the current soiling is a continuation of a preexisting problem.

Many metabolic or organ dysfunction diseases such as diabetes mellitus or Cushing's disease, renal disease or cystitis can lead to inappropriate urination by both dogs and cats. Similarly, gastrointestinal dysfunction or irritable bowel syndrome, pancreatitis or infections can lead to defaecation inside the home (Bain 2016, Gruen and Sherman 2016).

Pain can be a leading contributor in inappropriate soiling in senior pets. The inability to reach the litter box in time or go out to eliminate or pain while soiling can lead to avoidance of the area or the box. The knowledge or memory of pain during soiling may last even after the successful resolution of the problem. Sensory decline may mean that the pet simply cannot reach a designated soiling area.

Anxiety and fear contribute to inappropriate soiling. The pet may be afraid to return to a familiar elimination area due to another pet or threatening situation. Moreover, anxious animals may mark their environment to feel more secure inside. Cats may start soiling along the perimeter of the home or on owners' personal possessions (e.g. clothing, shoes or bed).

Changes in household routine or environment can also lead to inappropriate soiling. Owners may change jobs or shifts and, as a result, alter their dog's schedule. Senior dogs may find it difficult to adapt to changes and may start soiling at home. Moving litter boxes inside the home, changing litter type or box structure, preventing cats from going outdoors or renovating can all lead to avoidance of the litter box.

The first step is a full physical examination, blood analysis and urinalysis. Tests such as faecal sampling should be considered as well. Once medical problems have been ruled out, the emphasis can be placed on behavioural problems. A detailed history is the next step. Type of soiling (urine, faeces or both), frequency of soiling, locations, vertical vs. horizontal surfaces, quantity and time of day are only some of the details that should be noted. In addition, the pet's body language while soiling, other behavioural changes, possible triggers and changes to environment or household should be considered. The goal is to identify patterns and possible triggers. Soiling can be divided into two distinct behaviours: (a) inappropriate elimination that is a result of changes in routine, environment and learning new behaviours, which all can lead to aversion or preference of a location or litter or box type, and (b) marking behaviour that is a result of anxiety and the pet is soiling in order to feel more secure in its own environment (Pryor et al. 2001).

Pets with CDS will typically, but not always, soil without any particular pattern. Soiling will depend mostly on physiological need, for example, voiding urine when the bladder is full without relation to time of day or location (Landsberg and Araujo 2005).

2.1.1.6 Aggression

Aggression may appear in senior pets due to several reasons and indicate changes in relationships and interactions with owners and family, other pets at home and the environment. As with other problems, it is important to note if this aggression or predictors (behaviours that indicate pre-existence of covert aggression) existed in the past.

In most cases aggression is a result of fear or anxiety. While pets often prefer avoiding conflict, it may be that due to age-related changes they are limited and must choose to fight. Sensory reduction and reduced mobility are some of the reasons (see above).

Metabolic changes, organ dysfunction and neoplasia can all increase irritability, affect brain function and reduce the threshold for aggression. Therefore, a full physical exam and laboratory tests or diagnostic imaging (e.g. MRI) must be performed.

Learning may play a role in the development of aggression; the animal might be reinforced by the consequences of the behaviour. For example, aggression might be used to prevent children from approaching the pet; once successful in removing the threat, the animal may continue using aggression.

2.1.1.7 Repetitive Behaviours

Senior pets may start exhibiting repetitive behaviours such as pacing, licking themselves (usually one of the legs or flank) or rhythmic vocalisation (see above). These behaviours are usually ritualistic, excessive and evolve with time. This can lead to changes in activity levels and interactions with owners and the environment. In addition, avoidance of, or aggression towards, the owner may develop if the owner is trying to disrupt the pet from performing these behaviours.

Pain, neoplasia, organ dysfunction, metabolic diseases and hormonal abnormalities can all lead to the development of these behaviours. Anxiety and fear can also lead the pet to develop these behaviours initially as displacement behaviors, which may progress to stereotypic or compulsive behaviours (Luescher 2003).

Many owners believe that it is impossible to train and stimulate senior pets or that they should be left alone to rest due to joint disease or other painful conditions. While it is true that some adjustment might be required (reduced exercise, diet changes, rest, etc.), owners should still try engaging their pets. In some cases, lack of stimulation, or boredom, might be the cause of repetitive behaviours. Owners must ensure their pet has the possibility to practise normal species-specific behaviours even in old age.

The veterinary surgeon must first ensure the pet has normal outlets and a sufficient level of stimulation and interaction while considering possible limitations. Using feeding toys, chew toys, training and appropriate exercise all can help.

2.1.2 Medical Differentials of CDS in Dogs and Cats

Any disease process that can lead to signs of DISHAA (see Table 1.1) and other signs of cognitive dysfunction should be considered when assessing a senior pet. Diseases of the CNS can directly alter the pet's mentation, responsiveness and interactions. In addition, diseases outside the CNS can indirectly affect brain function (e.g. reduced oxygen perfusion, toxins and reduced glucose available for brain function) or cause other CDS-like signs. Moreover, chronic or debilitating disease can lead to increased anxiety, withdrawal or avoidance behaviour, increased irritability and aggression and house soiling (Table 2.3).

Table 2.3 Medical differentials of cognitive dysfunction in dogs and cats

System	Possible causes	Possible behavioural signs
Sensory	Cataracts/lenticular sclerosis	Fear/anxiety
	Loss of vision	Disorientation
	Loss of hearing	Decreased response to stimuli
		Reduced learning ability
		Aggression
		Avoidance
		Vocalisation
Pain/musculoskeletal	Degenerative diseases	Avoidance
	Arthritis	Reduced interest in exercise or play
	Muscular dystrophy	Altered response to stimuli; aggression
		Reduced self-hygiene
		Increased vocalisation
Cardiovascular	Mitral insufficiency	Disorientation
	Hypertension	Tiredness or reduced interest in play and activity
	Cardiomyopathy	Withdrawal/avoidance
		Irritability
		Fear/anxiety
		Changes in appetite
		Vocalisation
Endocrine	Diabetes mellitus	All signs of cognitive dysfunction
	Insulinoma	House soiling/urine marking
	Diabetes insipidus	Appetite—increased/decreased
	Hypothyroidism	Activity—increased/decreased/apathy
	Hyperthyroidism	Irritability
	Hyperadrenocorticism	Aggression
	Hypoadrenocorticism	Sleep-wake cycle
		Stereotypic—licking
		Restlessness—pacing
		Vocalisation
Digestive	Dental diseases	Reduced appetite
	Hepatic diseases	Aggression/irritability
	Infectious/inflammatory	Avoidance/withdrawal
	Constipation	House soiling
	Nutritional imbalances	Night-time waking
	Pain	Stereotypic—pacing/licking
		Coprophagia
Urinary	Renal diseases	House soiling/marking
	Urinary tract infection	Aggression
	Idiopathic cystitis	Withdrawal/avoidance
	Urolithiasis	Pacing
	Urinary incontinence	Sleep-wake changes

2.1.2.1 Sensory Decline

Age-related changes can affect hearing and vision of pets. Ageing of the lens may lead to nuclear sclerosis that may slightly affect the vision. In addition, age-related or other pathological causes, such as cataracts, can be partial or complete, leading to blindness. While dogs and cats rely mainly on olfaction, reduced vision and hearing can lead to changes in communication and interactions but not necessarily to increased aggression and anxiety (Farmer-Dougan et al. 2014). Reduced olfaction, while less common, is more difficult to cope with. Pets with sensory reduction are not able to appropriately evaluate the situation or environment in which they are found and have reduced ability to cope with it or with changes (Landsberg and Denenberg 2009).

2.1.2.2 Pain

Virtually any disease process that leads to pain will affect the pet's behaviour. Both acute and chronic pain will lead to increased irritability, withdrawal, changes in activity levels, reluctance to play and even aggression (Camps et al. 2012). Pain can also lead to increased anxiety and fear as the pet learns it cannot avoid threatening or painful triggers. The elimination of avoidance may lead to increased aggression.

While dogs are more likely to show overt signs of pain (e.g. limping, curling, rubbing or licking an area and vocalising), cats show more avoidance-related behaviour. For example, cats with degenerative joint disease may not show lameness but are more likely to avoid jumping on owners, playing or hunting (Landsberg and Denenberg 2009). Both dogs and cats with arthritis may refuse to interact with the owners or go for a walk. This may be viewed as changes in interactions. Animals with arthritis may develop a repetitive behaviour of licking the painful area (Frank 2014). The release of local endorphins, enkephalins and opioids may further reinforce and maintain this behaviour.

Oral or dental pain, as well as gastrointestinal pain, may lead to reduced appetite, lack of desire to interact or avoidance and possibly aggression when owners try petting the pet's head.

Pets with chronic pain may adjust to it by changing their habits and patterns of behaviour. These animals may eat less, have altered responses to various stimuli and withdraw from interactions. Pain can also reduce sleep duration and quality, and the animal may become more irritable.

Musculoskeletal diseases that lead to muscle weakness and wasting (e.g. muscular dystrophy, myositis and immune-mediated disease) can lead to pain, altered mobility and changes in interactions. Moreover, animals with these conditions may have altered responses to triggers, exhibit house soiling and be more irritable (Frank 2014).

2.1.2.3 Cardiovascular Disease

Compromised circulation and hypotension may lead to hypoxia in the brain. This in turn can lead to reduced awareness and social interactions and altered responses to stimuli. In addition, it can lead to tiredness, reduced activity and confusion.

Cats and dogs with cardiomyopathies have reduced appetite, exercise intolerance, tiredness and even increased irritability. These animals may appear more withdrawn and disoriented at times and have altered sleep-wake cycles (Landsberg 2005).

Advanced conditions such as chronic congestive heart failure and hypertension may lead to development of anxiety as a result of constant struggle with normal breathing, inability to exercise, taking frequent medications and a degree of hypoxia in the brain.

2.1.2.4 Endocrine Disease

Diabetes mellitus in both dogs and cats may lead first to an increase in appetite and later decrease. Animals with a ravenous appetite may show aggression over food and treats. Hyperglycaemia leads to excessive drinking and urination that in turn may lead to house soiling (Neilson 2004, Manteca 2011). Fatigue and lethargy are common signs of diabetes mellitus that can affect sleep-wake cycles. Advanced stages of diabetes can lead to hyperkalaemia, leading to withdrawal, reduced interactions, lethargy, disorientation and confusion.

Diabetes insipidus can lead to similar signs of house soiling, lethargy, irritability and aggression over water resources. Moreover, animals may have altered sleep-wake cycles, and night-time waking, searching for outlets to soil and drink.

Hyperadrenocorticism may mimic signs of chronic anxiety by inducing anxiety-like physical changes due to consistently high levels of cortisol in the body (activation of the hypothalamic–pituitary–adrenal axis). Animals may become more irritable and more aggressive. Increased appetite may be seen in early stages of the disease, together with polydipsia and polyuria. Many dogs with hyperadrenocorticism may exhibit exercise intolerance and avoid play, interactions and other physical activities (Bowen and Heath 2005).

Hypoadrenocorticism can lead to lethargy and apathy. In many patients hypoglycaemia can be noted as well as alterations of sodium and potassium. This will lead to withdrawal of the animal, reduced interactions, decreased appetite and changes in sleep-wake cycles. Exercise intolerance, disorientation and confusion are some of the signs seen during the low phase of the disease.

Hypothyroid, which is seen mostly in dogs (or iatrogenic in cats), will lead to slower metabolism. It can be seen as lethargy and changes in sleep-wake cycles, changes in appetite and reduced interactions with people and the environment. Dogs may refuse to go for a walk, avoid play and prefer to be left alone. Physical changes related to this condition (e.g. obesity, skin sensitivities and allergies and fatigue) may lead to increased irritability.

Hyperthyroid in cats (and iatrogenic in dogs) accelerates metabolism. Cats often appear to have a ravenous appetite, irritability and increased activity. This can lead to night-time waking, excessive vocalisation, aggression (over food or when interacting with other cats and people) and even repetitive behaviours (Neilson 2004).

While insulinoma is very infrequent in dogs and rare in cats, this tumour can certainly lead to behavioural changes also seen in CDS. The main result of over-secretion of insulin is hypoglycaemia. Dogs can be lethargic and withdrawn, avoiding interactions with people and the environment (Meleo and Peterson 2014). They may show more fear-related signs as they are not able to cope with changes. This will also lead to increased sleep and vocalisation and decreased activity. In advanced stages confusion and disorientation can be noted.

2.1.2.5 Gastrointestinal Disease

The feeling of nausea often accompanies many gastrointestinal diseases. Nauseous animals may show anxiety-related signs such as lip licking, excessive stretching and refusal to eat. Gastrointestinal pain can increase irritability, affect sleep leading to night-time waking and lead to avoidance of interactions.

Any disease that can increase intestinal motility and lead to diarrhoea (e.g. inflammatory bowel disease, infections and pancreatitis) may cause house soiling (Landsberg et al. 2013). In addition, pets that are well trained not to soil in the house may develop anxiety at times when the owners are not there to let them out. Moreover, these pets may pace repetitively, show night-time waking and excessive vocalisation and have increased irritability (Denenberg and Landsberg 2016).

Constipation may increase pain and irritability and lead to house soiling. Constipated pets may become more aggressive, avoid interactions and hide. If the need to soil exists overnight, sleep-wake cycles might be altered.

Hepatic diseases (e.g. hepatic insufficiency or failure, gall bladder inflammation or stones and hepatic neoplasias) may lead to pain and problems with digestion and absorption of nutrients. Dogs and cats may show avoidance and aggression and have reduced sleep. Furthermore, hepatic dysfunction may lead to toxaemia that affects the brain (hepatic encephalopathy) (Tisdall et al. 2000). This can have a direct effect on the animal's behaviour, including irritability, aggression, confusion and disorientation, and even seizures that may appear as displacement and repetitive behaviours (e.g. staring into mid-air, fly snapping and pacing).

Pancreatic diseases (such as pancreatitis, exocrine pancreatic insufficiency and tumours) can all cause pain, nausea and maldigestion. This, in turn, can lead to avoidance, increased irritability and displacement behaviours such as eating non-food objects (Becuwe-Bonnet et al. 2012). House soiling can occur as well.

2.1.2.6 Urinary System Disease

Renal insufficiency or failure is a common disease in senior cats and dogs. One of the earlier signs is polyuria and polydipsia. This can lead to house soiling in senior pets. In advanced stages uraemia can lead to nausea, pain and lethargy. These pets may appear anxious, irritable and disoriented. Reduced appetite, changes in interactions, sleep-wake cycle alterations and avoidance are all common signs of advanced renal disease (Landsberg and Denenberg 2009).

Urinary tract infection, idiopathic cystitis and feline urinary syndrome are all common reasons for house soiling, night-time waking and changes in interactions. Urinary stones (kidneys and bladder) can cause pain, irritability and aggression (Neilson 2004, Landsberg et al. 2013).

Urinary incontinence is not an infrequent problem in senior pets, especially neutered female dogs. It can lead to house soiling, increased irritability and problems with self-hygiene. Increased anxiety can be a consequence of the disease and a result of the owner's reactions.

Renal or bladder tumours can lead to house soiling, pain and avoidance, reduced activity, night-time waking, pacing and excessive vocalisation. In later stages confusion and disorientation may also be seen.

2.1.3 Neurological Differentials of CDS in Dogs and Cats

A systematic diagnostic approach is of paramount importance in the face of non-specific clinical signs. As seen in the first part of this chapter, there is a large overlap in the clinical signs seen with CDS and behavioural, metabolic, neurological or painful conditions. One cannot stress enough the importance of the signalment, history taking, physical and neurological examinations to form a problem and then differential list before pursuing advanced laboratory or imaging testing. Here is the step-by-step approach the authors use in patients presented for signs that could be suggestive of CDS.

2.2 Diagnostic Approach to Reach a Presumptive Diagnosis of CDS

2.2.1 Signalment

Signalment will have an important impact on the organisation of the differential diagnosis list. Cats or dogs younger than 6 years of age are very unlikely to be affected by CDS—at least not at a clinically detectable level. While the individual's breed seems less relevant, it should be incorporated in the clinical reasoning (Head et al. 2009a, b). The clinician should be familiar with breed predispositions (Table 2.4) as these could have an impact on the suspicion of a non-infectious inflammatory brain disease, a neoplastic process, an infectious process or a metabolic condition. Some databases are available online or in veterinary textbooks to inform the clinician about disease prevalence for specific breeds.

2.2.2 History Taking

Thorough history serves to identify indicators of any neurological or metabolic diseases that would lead you to reject CDS as your suspected diagnosis (e.g. the presence of polyuria or polydipsia (PU/PD), recurrent urinary tract infection, seizures, circling and head pressing). Focus the questions around the DISHAA questionnaire (see Table 1.1).

Table 2.4 Suspected breed predisposition

Conditions	Breeds
Gliomas	Boxer, French bulldog, Boston terrier
Choroid plexus tumours	Golden retriever, Dalmatian, Irish setter
Meningiomas	Golden retriever, miniature schnauzer, rat terrier
Non-infectious meningoencephalitis	Pug, maltese, Yorkshire terrier, West Highland white terrier, rottweiler
Ceroid lipofuscinosis	Tibetan terrier, dachshund
L2-HGA	Yorkshire terrier, Staffordshire bull terrier, West Highland white terrier
Vitamin E deficiency	English cocker spaniel
Congenital PSS	Yorkshire terrier
Ischaemic strokes	Greyhound, Cavalier King Charles spaniel
FIP infection	Purebred cats

L2-HGA L2-hydroxyglutaric aciduria, *PSS* Portosystemic shunt, *FIP* Feline infectious peritonitis

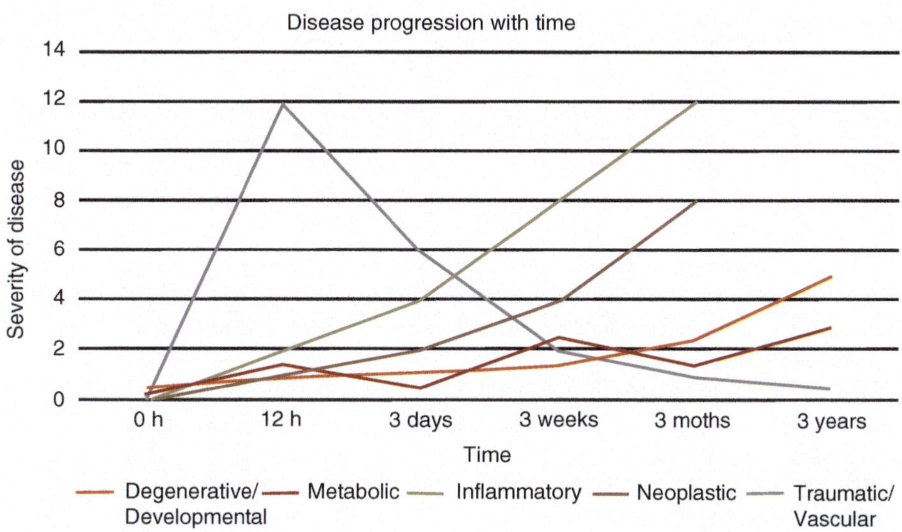

Fig. 2.1 Temporal progression of neurological diseases

Another important consideration is the onset and progression of clinical signs, which is useful in forming different lists (Fig. 2.1). As mentioned, CDS should have an insidious onset and is slowly progressive in nature; if the animal history suggests differently, the diagnosis of CDS is not likely.

2.2.2.1 Disorientation

Is this occurring at any particular time of the day? Is it happening during exercise, in the middle of the day or immediately after feeding? The former could raise the suspicion towards hypoglycaemia, while the latter can be seen in animals with hepatic insufficiency following a large protein intake. If disorientation occurs mainly or solely later in the evenings or in dim lighting conditions, then the clinician should be suspicious of an altered sense of vision (e.g. nyctalopia seen with progressive retinal atrophy or with retinal degeneration in patients suffering from neuronal ceroid lipofuscinosis).

2.2.2.2 Alterations in Interactions with Owners, Other Pets and the Environment

This could occur in specific situations: a pet who growls, hisses or bites when woken could have reduced hearing, while an animal reluctant to play with other animals or to go for walks could be in pain or weak. Clinical signs that include compulsive pacing, circling to the same side, head pressing against walls or objects, hemineglect syndrome, episodes of opisthotonus or seizures would suggest a metabolic disorder or intracranial lesion.

2.2.2.3 House Soiling

House soiling can be seen in a variety of situations, but history will provide invaluable information to guide the clinician in their diagnostic approach. The presence of polyuria, polydipsia (water intake >100 mL/kg/day), pollakiuria, periuria or dysuria should be documented and will require further, more specific investigation. The diagnostic approach for an animal with polyuria and polydipsia would certainly be different from that for an animal accidentally passing faeces inside the house. Intestinal parasites control should also be documented as well as the presence of chronic gastrointestinal signs (vomiting, diarrhoea, dyschezia, regurgitation, melaena, weight loss, etc.) (Head et al. 2009a, b). The clinician should also ask whether the animal seems to be conscious of the fact they are passing faeces or urine and the frequency and nature of the faeces/urine. For example, being aware of urination and the urine being passed with pollakiuria and haematuria would suggest a lower urinary tract disease.

Differential diagnosis for house soiling: polyuria, polydipsia, urinary tract infection, urinary incontinence, faecal incontinence and loss of training, and painful conditions.

2.2.2.4 Changes in Activity

Again the clinician should focus the questioning on the possibility of the presence of a hypometabolic state or hypermetabolic state. Any recent weight loss or gain should be documented as well as the presence of excessive panting (hypertension, cardiorespiratory disease) or polyphagia (diabetes mellitus, hyperadrenocorticism, hyperthyroidism, acromegaly).

2.2.2.5 Establishment of the Most Relevant Complaint(s)

This step ensures the clinician lists specific and less specific complaints but also attends to the needs and expectations of the pet's owner, irrespective of the animal's problem. Managing the owner's expectations means taking into account the historical and clinical findings but also addressing the owners' concerns. CDS could be the main disease or could be concomitant with other conditions.

The greater the number of DISHAA signs (see Table 1.1) present, the stronger the suspicion for CDS, although concurrent or alternative differentials remain possible.

2.2.3 Patient Examination

2.2.3.1 Physical Examination

The physical examination should be thorough and systematic (Table 2.5). An in-depth description of the physical examination is beyond the scope of this chapter but should focus on the detection of painful body areas or signs suggestive of an underlying metabolic disease (ulcerative tongue lesions and small irregular kidneys could be part of a uremic syndrome, organomegaly could suggest acromegaly, etc.). A cat's neck should be thoroughly palpated for the presence of a thyroid slip that could be indicative of an enlarged thyroid gland.

2.2.3.2 Orthopaedic Examination

Due to the aged population being considered, concomitant orthopaedic issues are likely. The clinician will then be faced with the dilemma of determining whether the abnormalities found on orthopaedic examination are concomitant/incidental or the primary cause for the presenting complaint. In particular, osteoarthritis is common in ageing cats and dogs and could be a plausible reason for reduced activity, fearful behaviour or altered interactions. It is worth noting that generally speaking scuffing of the nails is associated with a neurological rather than an orthopaedic complaint.

Joint or orthopaedic conditions seen in older patients include osteoarthritis, cranial cruciate ligament injury, intervertebral disc disease, bone neoplasia and discospondylitis.

Table 2.5 Physical examination findings suggestive of a neurological dysfunction

Clinical signs	Potential significance
Head tilt	Brainstem lesion, otitis, idiopathic vestibular syndrome, ear neoplasia
Head turn	Contralateral forebrain lesion
Scuffed nails	Proprioceptive deficits
Facial twitching	Partial seizures, myokymia
Self-mutilation	Neuropathy
Anisocoria	Ocular lesion, Central or peripheral nervous system lesion
Circling	Contralateral forebrain lesion
Head pressing	Raised intracranial pressure

2.2.3.3 Ophthalmological Examination

The goal of a thorough ophthalmological examination is to detect any vision loss but also to document further clues for the diagnostic puzzle. The clinician should assess the presence of cataracts, retinal haemorrhage, retinal degeneration, retinal detachment or chorioretinitis. Retinal haemorrhage and chorioretinitis are more specific findings and would warrant further investigation that may lead to the more efficient diagnosis of a primary underlying process. The optic nerve will be visible in most animals and should be assessed for the presence of optic neuritis or papill-oedema. Abnormal vision in the absence of ophthalmological abnormality will be strongly suggestive of a central nervous system lesion.

2.2.3.4 Neurological Examination

The neurological examination allows you to assess if the patient has a neurological problem and to localise the lesion. Localisation in turn, when used in combination with signalment and history, allows the creation of a differential list which will guide your diagnostic plan and let you know if CDS should be considered a differential.

The neurological exam is completed by (a) observing the neurological patient and (b) performing specific hands-on tests. The patient is assessed in five major areas (Table 2.6)

Following this, an assimilation of the information allows you to localise to one (or more) of eight major anatomical regions: forebrain, brainstem, vestibular-cerebellar system, spinal cord C1-C5, C6-T2, T3-L3 and L4-S3, and neuromuscular system.

It is important to perform a full neurological exam in a patient you suspect with CDS as any abnormality outside changes of behaviour or mentation may suggest an alternative diagnosis. This could be changes suggestive of a forebrain disease or a multifocal exam which could suggest raised intracranial pressure resulting in CNS herniation, diffuse degenerative process, a multifocal inflammatory/infectious disease or neoplasia.

It is worth considering that some older dogs could have concurrent diseases (such as some signs of an L4-S3 localisation associated with lumbosacral disease) alongside forebrain signs of CDS, but a multifocal exam with a progressive history could also be highly suggestive of neoplastic and inflammatory/infectious diseases, so it is worth trying to exclude these as possibilities both clinically and with diagnostics testing when felt appropriate (Head 2009a, b).

Table 2.6 Areas being tested and methods of assessment

Area of testing	Method of testing
Behaviour and consciousness	Observation
Cranial nerves	Observation and the hands-on examination
Gait and posture	Observation
Postural reactions	Hands-on examination
Spinal reflexes	Hands-on examination

Describing how to perform a full neurological exam is beyond the scope of this article and can be found in dedicated textbooks. We will now discuss changes that, when found, would lead you away from considering CDS as a diagnosis. Doing a full neurological examination takes time so ensure you leave more than the 5–15 minutes of a general consultation.

2.2.3.5 Changes on Neurological Examination that Would Not Be Expected with CDS

Items in bold are in the authors' experience the most common changes seen in cases with a purely forebrain (FB) localisation that would suggest CDS is not a differential. Although all items of the neurological exam are to be checked, we have tried to place them in order of preference/importance to screen the neurological system for disease. For example, looking for nystagmus (in particular a positional nystagmus) is a good screen of brainstem function given the extensive nature of the vestibular nuclei in this region of the brain. This is obviously important if inflammatory diseases are a differential or if you are worried about raised intracranial pressure where nystagmus may be encountered secondary to brain herniation. Items with a FB in brackets are tests that may be abnormal with a purely FB lesion.

Behaviour and consciousness:

- **Seizures (FB)**
- **Persistent circling (FB)**
- **Head pressing (FB)**
- Stuporous mentation (FB)
- Comatose mentation
- Hemineglect syndrome (FB)

Notes: Persistent circling can be both secondary to a forebrain lesion and a vestibular-cerebellar disturbance, so you would want to perform tests on the vestibular-cerebellar system to differentiate the two. Circling due to a forebrain lesion tends to be in less tight circles. Stuporous and comatose mentation is rarely seen with forebrain diseases in the absence of numerous other forebrain deficits on exam as they would need to affect such diffuse areas of brain tissue. As a result, it is more common to see these with brainstem disease. The exception would be a disease affecting the diencephalon of the forebrain (e.g. a pituitary neoplasm).

Cranial nerves testing:

- **Reduced/absent menace response (FB)**
- **Reduced/absent visual tracking (FB)**
- **Reduced/absent facial sensation usually found with reduced/absent nasal mucosal response (FB)**
- Nystagmus-spontaneous or positional
- Anisocoria (FB)
- Facial droop/paresis (FB)
- Reduced palpebral light reflex

- Strabismus
- Absent/reduced vestibulo-ocular reflex
- Reduced gag reflex
- Reduced/absent corneal reflex
- Reduced/absent dazzle reflex
- Masticatory muscle wastage
- Inability to close the jaw

Notes: The menace response and positional nystagmus are considered by the authors the most important cranial nerve tests to screen the forebrain and brainstem, respectively. A head tilt and facial paresis are both very rare with forebrain disease but remain possible. It is worth noting that finding these signs in isolation without other changes suggestive of a forebrain localisation would point towards alternative localisations in the neurological system.

Gait and posture:

- **Deficits in postural reactions (FB)**
- **A head turn (FB)**
- **Circling (FB)**
- Paresis/ataxia (FB)
- A head tilt (FB)
- Decerebrate posture
- Decerebellate posture

Notes: Deficits in postural reactions can be detected through paw placement, hopping, tactile placement, visual placement, wheelbarrowing and hemi-walking. A head tilt is very rarely seen secondary to a purely forebrain disease and usually only occurs secondary to vascular lesions affecting the thalamus.

Relevant spinal reflexes:

- Patellar reflex
- Withdrawal reflex (thoracic and pelvic limbs)
- Perianal reflex/anal tone
- Cutaneous trunci

Notes: No spinal reflexes should be abnormal with CDS or any other forebrain, cerebellar-vestibular or brainstem disease. I have only included the spinal reflexes that are reliable, but others do exist. When assessing these, we would also consider muscle wastage, tone, a crossed extensor and evidence of spinal pain.

2.2.4 Problem List and Differential Diagnosis

2.2.4.1 Problem List

After the history has been taken and after the patient's examination has been performed, a problem list is established. The clinical suspicion of CDS will increase with each addition of the DISHAA (see Table 1.1) clinical sign. The problem list will also help the clinician by listing problems that are more specific (PU/PD vs. lethargy, neurological deficits vs. lethargy) or that are more likely to lead to the diagnosis of an underlying alternative pathology (seizures vs. increased fearful behaviour).

2.2.4.2 Differential List

Using the problem list and then considering the signalment, history, and physical and neurological examinations will allow formation of a differential list. We have compiled a differential list below based on conditions with overlapping signs with CDS in patients over the age of 6 years.

2.2.4.3 Differential Diagnosis: DAMNITV

Here is a non-exhaustive list of the conditions that may enter the differential diagnosis list of a clinician faced with a patient suffering from clinical signs that could be due to a metabolic disease, neurological disease or CDS. The authors use the DAMNITV (degenerative, anomalous, metabolic, nutritional or neoplastic, inflammatory or infectious, traumatic or toxic, vascular) or VITAMIND algorithm.

2.2.4.4 Degenerative Differentials

Ceroid Lipofuscinosis

Aetiopathogenesis Ceroid lipofuscinosis is a neurodegenerative disease characterised by the accumulation of autofluorescent lipopigments (ceroid- or lipofuscin-like lipopigments) in neurons and other cells within the body. While most affected dogs and cats will present with signs early in life (within the first 2 years), ceroid lipofuscinosis is one of the few storage diseases that can present in adulthood. This is particularly true in the Tibetan terrier and dachshund breeds who may develop obvious signs when 6 years or older and can survive past 10 years of age. The pathophysiology of the disease is yet to be elucidated as it is unclear whether the mechanism of injury is a direct consequence of the pigment accumulation in cells or if it involves an abnormal mitochondrial function. The advances of molecular genetics have permitted the identification of multiple gene defects causing enzymatic dysfunction, leading to ceroid lipofuscinosis in specific dog breeds.

Clinical Signs The clinical signs reflect the organ dysfunction where the pigment accumulates. The two main systems affected are the visual system and the forebrain. The owners may report a progressive decline in vision with a tendency to bump into objects, especially at night (nyctalopia). Behavioural changes include loss of training, development of unprovoked aggression, progressive clumsiness, pacing, difficulty recognising the owners and house soiling. Affected dogs can also be easily startled and can develop overwhelming anxiety. Neurological deficits are usually symmetrical and may be initially subtle or absent. Vestibular signs with an intermittent loss of balance, decreased vision and mild proprioceptive deficits in all limbs despite normal spinal reflexes can progress later in the disease to signs of head pressing or severe ataxia. Ophthalmological examination may reveal changes consistent with retinal degeneration (tapetal hyperreflectivity, thinning and paucity of the retinal arteries).

Clinical signs that overlap with CDS:

- Loss of training
- Unprovoked aggression
- Progressive clumsiness
- Pacing
- Difficulty recognising the owners and house soiling

Clinical signs that would not overlap with CDS:

- Loss of vision
- Ataxia
- Neurological deficits

Diagnosis/Exclusion For breeds in which a known causative mutation has been identified, a diagnosis can be reached with genetic tests performed on DNA samples (EDTA blood sample or buccal swab). The diagnosis can also be supported by advanced imaging, with MRI changes of generalised brain atrophy despite normal spinal fluid results being suggestive of a neurodegenerative disease. Sometimes EEG changes will support this disease process too. Contrast enhancement of the meninges has also been reported in a cohort of affected chihuahuas.

Treatment Unfortunately, there is no known treatment for this condition. The progression is variable, but some dogs will survive for years before euthanasia is elected based on welfare or security grounds (in the case of unprovoked aggression unresponsive to treatment).

L-2-Hydroxyglutaric Aciduria

Aetiopathogenesis L-2-hydroxyglutaric aciduria (L2-HGA) is a neurometabolic disease with an unknown pathogenesis. It is suggested that L2-HGA may act as a direct toxin on the central nervous system or act through oxidative stress or mitochondrial dysfunction.

Clinical Signs This disease has only been reported in the Staffordshire bull terrier (SBT), West Highland white terrier and Yorkshire terrier to date. While L2-HGA is usually reported in younger animals, some dogs have first presented with the disease after 7 years of age.

Clinical signs are progressive and include behavioural changes (including inability to learn new tasks and forgetting learned tasks, getting stuck under tables and chairs/corners, etc.), changes in mentation (e.g. obtundation), head pressing as well as psychomotor retardation, dysmetria, cerebellar signs (including dysmetria, head tremors and reduced menace response in the presence of normal vision), muscular stiffness after exercise, ataxia, tremors and seizures.

Clinical signs that overlap with CDS:

- Obtundation and behaviour changes

Clinical signs that would not overlap with CDS:

- Head pressing
- Seizures
- Ataxia
- Dysmetria
- Cerebellar signs
- Muscular stiffness after exercise and tremors

Diagnosis/Exclusion Diagnosis is made based on consistent clinical signs in combination with consistent laboratory findings, with or without imaging studies and genetic testing. In particular, demonstrating the presence of increased levels of L2-HGA in urine or serum with metabolic testing is diagnostic. In SBT there is a genetic test available, and MRI imaging of the central nervous system is characteristic.

Treatment There is no specific treatment. In people, various dietary supplements have shown anecdotal improvements.

Other Neurodegenerative Diseases

Most neurodegenerative disease will present early in life and will be considered as unlikely differentials for CDS.

Lafora disease can affect older dogs, in particular wire-haired dachshunds, beagles and basset hounds, but most dogs will present with myoclonic seizures, although some behavioural changes may be reported. The diagnosis is made by exclusion of metabolic and intracranial lesions and identification of the known mutation causing Lafora disease, when available. Tissue biopsy may allow the identification of polyglucosan bodies within the brain, muscles, skin or liver. Treatment is symptomatic to control the seizures.

Neuroaxonal dystrophy is another neurodegenerative disease for which the chief complaint could be behavioural changes, although the neurological examination will almost invariably reveal neurological deficits suggestive of a cerebello-vestibular dysfunction. The aetiopathogenesis is poorly understood, but an abnormal metabolism of vitamin E is often cited as a possible cause. Diagnosis can be supported by an abnormal MRI scan, although post-mortem examination is often necessary to reach a definitive diagnosis. The benefit of vitamin E supplementation is yet to be proven.

2.2.4.5 Anomalous: Brain Malformations

Aetiopathogenesis Brain malformations occur due to a failure of development of normal brain tissue or its destruction during foetal or embryonic life. The latter can be seen secondary to infectious, traumatic or vascular injury to the developing brain. While most malformations would be expected to cause problems in young animals, some malformations may remain silent and may manifest only later as the malformation gets bigger (e.g. quadrigeminal cysts, hydrocephalus) or causes obstruction of the spinal fluid pathways (e.g. fourth ventricle arachnoid cysts). The injury results from direct pressure on the brain tissue or due to secondary increased intracranial pressure following the development of obstructive hydrocephalus.

Clinical Signs The clinical signs reflect the localisation of the lesion. With quadrigeminal cysts and fourth ventricle cysts, compression of the brainstem may lead to signs of central vestibular syndrome, including an obtunded mental status, lethargy, and neurological deficits such as a head tilt, nystagmus and proprioceptive deficits. An animal with increased intracranial pressure can manifest signs of neck pain and, as a consequence, may become aggressive and withdrawn.

Clinical signs that overlap with CDS:

- Altered mental status
- Aggression

Clinical signs that would not overlap with CDS:

- Head tilt
- Nystagmus
- Proprioceptive deficits
- Ataxia, circling
- Seizures
- Pain

Diagnosis/Exclusion The diagnosis is made by advanced imaging, with an MRI scan of the brain characterising the lesions and their potential consequences as oedema, mass effect, brain herniation or syringohydromyelia. One dilemma resides in the fact that some animals can have incidental cystic lesions as quadrigeminal cyst not causing clinical signs but diagnosed later in life for the investigation of a clinical complaint of lethargy or abnormal behaviour. In this instance the clinician has to use his judgement to decide whether or not the cyst can explain the clinical complaint and the clinical signs of the animal and whether or not the cyst should be treated. One retrospective paper has suggested that the relative size and compression of quadrigeminal cysts may relate to incidence of clinical signs.

Treatment The treatment of brain malformations, in particular cystic lesions, can be medical, with drugs used to reduce oedema or Cerebrospinal fluid (CSF) production, or surgical, to remove the lesion or restore a CSF pathway to resolve obstructive hydrocephalus.

2.2.4.6 Metabolic Differentials

Hypoglycaemia

Aetiopathogenesis Hypoglycaemia can be caused by hyperinsulinaemia (e.g. insulinomas), Addison's disease, hepatic failure, sepsis, neoplasia, repeated inappropriate medication dosing and rarely hunting dog hypoglycaemia or starvation. Other general causes of hypoglycaemia include toy breed hypoglycaemia and xylitol or ethylene glycol toxicity. The lack of glucose results in abnormal neuronal function and activation of the sympathetic nervous system, resulting in clinical signs.

Clinical Signs Any value lower than the reference interval should be considered suspicious for hypoglycaemia, but signs do not usually become apparent until <2.5 mmol/L. Hypoglycaemia can be seen in any age and breed of dog, but xylitol or ethylene glycol toxicity should have a peracute/acute onset of signs, and toy breed hypoglycaemia would be seen in the young.

Signs of hypoglycaemia include lethargy, tremors/muscle facilitations, weakness, changes in behaviour (nervousness, vocalisation, irritability, forgetting learned behaviours), tachycardia, visual disturbances, hypothermia, changes in mentation (obtundation, stupor or comatose state) and seizures. Given the list of differentials above, some cases may show other systemic signs such as polyphagia and vomiting.

Clinical signs that overlap with CDS:

- Lethargy
- Obtundation and behaviour changes

Clinical signs that would not overlap with CDS:

- Seizures
- Blindness
- Hypothermia and tachycardia
- Weakness or tremors

Diagnosis/Exclusion Diagnosis is made based on consistent clinical signs in the presence of recorded hypoglycaemia. It is necessary to perform a starved glucose measurement to exclude the disease, while inclusion would be any signs of hypoglycaemia (defined as <3.3 mmol/L). The authors perform serial blood glucose levels over a 24-hour starvation period and may consider a serum fructosamine to exclude hypoglycaemia as a cause for encephalopathy.

Treatment In the acute setting, this revolves around feeding, intravenous dextrose and glucose-infused fluid therapy. The underlying cause should then be investigated and treated appropriately.

Hepatic Encephalopathy

Aetiopathogenesis Hepatic encephalopathy (HE) can occur due to portosystemic shunting (PSS), which can be acquired or congenital, and due to acute or chronic hepatopathies, for example, feline hepatic lipidosis. Cases with PSS or microvascular dysplasia may present later in life with signs of HE secondary to a trigger such as infection, constipation, diarrhoea, increased dietary protein, gastrointestinal haemorrhage, electrolyte imbalances and arginine deficiency in cats with hepatic lipidosis (Holt et al. 2002).

A number of proposed mechanisms have emerged to explain the pathogenesis of hepatic encephalopathy. These revolve around increased levels of ammonia (which may be worsened by inflammation), increased levels of neurosteroids, increased manganese, increased oxidative stress through upregulated reactive oxygen and nitrogen species, and decreased aromatic amino acids. Ultimately, a dysfunction in neurotransmitters, neurotransmitter receptors and transporters, alteration in blood-brain barrier and electrolytes and alterations in neuronal and glial cell function or survival result in encephalopathic clinical signs (Wolschrijn et al. 2011).

Clinical Signs HE can be seen in dogs of any age and any breed. PSS tends to be seen in younger dogs, while acquired hepatopathies tend to be seen in older patients.

HE signs range from mild (e.g. mild obtundation) to more severe (e.g. seizures or comatose state) and often have an episodic nature in history often worsened by a meal. In a study of 80 dogs with hepatic encephalopathy, the most common historical signs were obtundation (33%), altered behaviour (29%), head pressing (28%), ataxia (26%), seizures (24%), vomiting (24%), lethargy (24%), ptyalism (23%) and blindness (19%). Other signs include anorexia, weight loss, diarrhoea, polydipsia and polyuria or poor weight gain. On neurological examination the most common findings were obtundation (31%), ataxia (20%), weakness (10%), proprioceptive deficits (9%), seizures (8%), circling (6%), cranial nerve deficits (5%), stupor (5%) and tremors (4%).

Clinical signs that overlap with CDS:

- Lethargy
- Obtundation and behaviour changes

Clinical signs that would not overlap with CDS:

- Head pressing
- Seizures
- Vomiting
- Diarrhoea
- Weight loss
- Icterus
- Organomegaly
- Ptyalism
- Blindness
- Circling
- Ataxia
- Neurological deficits
- Tremors
- Stupor

Diagnosis/Exclusion Diagnosis is made on consistent clinical signs, laboratory findings and imaging studies followed by response to treatment (in the exclusion of other causes of encephalopathy).

A complete blood count and biochemistry profile (including urinalysis) may be normal or may show a non-regenerative, microcytic anaemia, hypoglycaemia, hypoalbuminaemia, hypocholestrolaemia, increased Alkaline phosphatase (ALP) or Alanine aminotransferase (ALT), hyperbilirubinaemia, decreased urea and decreased urine concentrating ability with ammonium biurate crystals on cytology of urine.

Other laboratory tests for liver dysfunction include analysis of ammonia, resting bile acids or more importantly a dynamic bile acid stimulation test. Fasting ammonia levels have been shown to be relatively sensitive for detecting PSS with 91% in dogs (>59 μmol/L) and 83% (>94 μmol/L) in cats, with fasting bile acids being 78% for dogs (>58 μmol/L) and 100% in cats (>34 μmol/L). Although the sensitivity and specificity are not known for dynamic bile acids for hepatic dysfunction, this is proposed to be the superior test compared to a resting bile acid.

A definitive diagnosis of hepatic dysfunction causing encephalopathy must come from either demonstrating a shunting vessel with CT with IV contrast, mesenteric portography and abdominal ultrasound or less commonly scintigraphy for a PSS, or through imaging and liver biopsy for other hepatopathies.

Treatment This revolves around medical and surgical therapy. Medical therapy includes enemas in the acute setting alongside non-absorbable disaccharides (e.g. lactulose), antibiosis and high-quality protein diets. Anticonvulsants should be used if seizures are occurring. Other proposed supportive drug therapies include flumazenil (as an antagonist of benzodiazepines), probiotics and prebiotics, L-carnitine supplementation and L-ornithine-L-aspartate. Addressing and treating the underlying cause for the hepatic disease should also be undertaken.

Renal Encephalopathy

Aetiopathogenesis Renal encephalopathy (RE) comes secondary to renal failure. Renal failure can be caused by many differing disease processes—the scope of which is beyond this book chapter. These can be referenced in various medical textbooks (Holt et al. 2002).

There are many proposed mechanisms by which RE may come about and include increased parathyroid hormone levels, ionic imbalances with associated hyperosmolality, hypertension, uraemia and acidosis. These cause alterations in neuronal and glial cell function or survival by altering neurotransmitters, neurotransmitter receptors and transporters, the blood-brain barrier, cell membrane stability and ischaemia by various means.

Clinical Signs RE can be seen in any age or breed but is very rare in the author's experience. First and foremost these patients will present with signs of renal failure—polydipsia, polyuria, inappetence, weight loss, vomiting, lethargy, etc. In addition to the above signs, RE can cause changes in mentation (obtundation progressing to coma), seizures, muscle tremors, generalised weakness and irregular patterns of respiration.

Clinical signs that overlap with CDS:

- Lethargy
- Obtundation
- House soiling

Clinical signs that would not overlap with CDS:

- Stupor or comatose state
- Seizures
- Irregular patterns of respiration
- Muscle tremors
- Polydipsia
- Polyuria
- Weight loss
- Vomiting
- Inappetence

Diagnosis/Exclusion Diagnosis is made through consistent clinical signs and supporting laboratory findings. A complete blood count and biochemistry panel include full urinalysis and blood pressure evaluation.

Electrolytes Disturbances

Aetiopathogenesis Electrolytes are indispensable for the normal function of living cells by allowing the creation of electric potentials and message transduction. Potassium, sodium, calcium and magnesium are the most common ions inside or outside cells, and their disturbances will affect cell and organ function, with the central or peripheral nervous system being particularly affected. In older animals, the two electrolytes disturbances of interest are sodium or calcium abnormalities (Attems 2005).

Clinical Signs Hypernatraemia can cause systemic hypertension with the clinical signs mentioned earlier in this chapter. As an acute change, it can cause a marked demyelination (as below) and in a chronic state when corrected quickly can lead to cytotoxic oedema secondary to the formation of idiogenic osmoles within the neurons. Chronic hyponatraemia is unlikely to cause clinical signs on its own, but rapid correction of hyponatraemia (more rapidly than 0.5 mmol/L/h) can lead to signs of pontine and extrapontine demyelination including altered mental status, disorientation, seizure and proprioceptive deficits suggestive of a brainstem or forebrain lesion.

These signs can persist despite resolution of the electrolyte imbalance (Yoshino et al. 1996). Hypocalcaemia will mainly cause tremors, cramping, facial rubbing, panting, behavioural changes (disorientation, restlessness, excitation, aggression, hypersensitivity to stimuli), lethargy and cardiac arrhythmias. Hypercalcaemia can cause polyuria, polydipsia (which can lead to house soiling), anorexia, lethargy, weakness, dehydration, seizures, arrhythmias, constipation and renal failure.

Clinical signs that overlap with CDS:

- Behavioural changes
- Disorientation
- Lethargy
- House soiling

Clinical signs that would not overlap with CDS:

- Muscle cramps
- Arrhythmia
- Neurological deficits
- Seizures
- Arrhythmias

Differential diagnosis for hypercalcaemia:

- Hyperparathyroidism
- Hypercalcaemia of malignancy
- Hypervitaminosis D
- Hypoadrenocorticism
- Renal insufficiency
- Osteolytic lesions
- Granulomatous diseases
- Sepsis
- Laboratory error
- Idiopathic (cats)

Differential diagnosis for hypocalcaemia:

- Hypoparathyroidism
- Eclampsia
- Hypovitaminosis D
- Hyperadrenocorticism
- Renal insufficiency
- Pancreatitis
- Laboratory error

Differential diagnosis for hypernatraemia:

- Pure water deficit: hypodipsia, diabetes, insipidus fever
- Hypotonic fluid loss: gastrointestinal loss, renal losses, third space losses, burns
- Impermeant solute gain: salt poisoning, hypertonic fluid administration, hyperaldosteronism, hyperadrenocorticism

Differential diagnosis for hyponatraemia:

- With hypervolemia: cardiac failure, advanced renal failure, severe liver disease
- With normovolaemia: psychogenic polydipsia, syndrome of inappropriate antidiuretic hormone secretion, myxoedema coma of hypothyroidism
- With hypovolemia: gastrointestinal losses, third space losses, burns, hypoadrenocorticism

Diagnosis/Exclusion The diagnosis is made by measuring blood electrolytes. Total calcium being dependent on blood pH, protein concentration and other parameters, it is not a reliable marker of the available body calcium; ionised calcium should be performed to rule out hypo- or hypercalcaemia.

Treatment The treatment will aim at correcting the electrolytes disorder but most importantly its underlying cause. The correction of hypernatraemia or hyponatraemia should be done slowly, with the sodium being decreased or increased by no more than 0.5 mmol/L/h to avoid the development of brain oedema or pontine and extrapontine demyelination syndrome, respectively. Calcium correction should be done while the ECG of the animal is monitored for arrhythmias.

2.2.4.7 Nutritional

The origin of vitamin deficiency in dogs and cats is an inappropriate dietary intake, a lack of absorption or the presence of molecules that inhibits or destroys the vitamins in the food. The two vitamins of interest to us are vitamin E and vitamin B1 (thiamine) (Head et al. 2009a, b).

Vitamin B1 Deficiency

Aetiopathogenesis Vitamin B1 (thiamine) is a coenzyme in the Krebs cycle and additional enzymatic pathways involved in energy production. The absence of thiamine may lead to damage of high-energy structures, in particular brainstem nuclei in dogs and cats but also the heart and retina.

Clinical Signs The clinical signs seen in dogs or cats suffering from thiamine deficiency include abnormal mentation, ventro-flexion of the neck/head, blindness, vestibular ataxia, head tilt, nystagmus, mydriasis, facial paralysis and seizures (Gold et al. 2015).

Clinical signs that overlap with CDS:

- Abnormal mentation

Clinical signs that would not overlap with CDS:

- Ventro-flexion of the neck/head
- Ataxia, head tilt, nystagmus
- Facial paresis, blindness
- Seizures (Head et al. 2009a, b)

Diagnosis/Exclusion Thiamine deficiency is diagnosed by measuring total thiamine level in blood, but the diagnosis can also be strengthened by the presence of characteristic bilaterally symmetrical lesions in brainstem nuclei on MRI and by an abnormal organic acid profile in urine.

Treatment The treatment involves oral or parenteral supplementation of the deficient vitamin and correction of the underlying cause. Animals suffering from thiamine deficiency can recover fully.

Vitamin E Deficiency

Aetiopathogenesis Vitamin E is a potent antioxidant whose deficiency leads to oxidative stress injury, this being pronounced in the muscles, CNS and retinas. Vitamin E deficiency is suspected to have a genetic basis in English cocker spaniels (Norton et al. 2016).

Clinical Signs The clinical signs associated with vitamin E deficiency include progressive vision loss (retinal epithelial pigment dystrophy), reduced activity and reluctance to walk, muscle weakness and neurological deficits consistent with a myopathy or a cerebello-vestibular localisation.

Clinical signs that overlap with CDS:

- Reluctance to exercise or walk

Clinical signs that would not overlap with CDS:

- Neurological deficits, blindness

Diagnosis/Exclusion The diagnosis of vitamin E deficiency is made by measuring serum vitamin E concentration and by ruling out other extracranial and intracranial causes. MRI and CSF analysis have been reported as normal in dogs for which it has been performed.

Treatment Cocker spaniels affected by vitamin E deficiency supplemented orally did not experience an improvement in their visual function, but their neurological function improved or stabilised, in some cases for years (Johnson et al. 2015).

2.2.4.8 Neoplastic Brain Disease

Aetiopathogenesis Intracranial neoplasia can be primary if originating from the brain structures themselves, secondary if invading the brain by local extension (e.g. from a nasal, ocular or skull origin) or metastatic if affecting the brain by vascular spread of a distant neoplastic process. In dogs and cats, primary intracranial tumours are more common, while metastatic neoplasias (hemangiosarcoma, carcinomas or multicentric lymphoma) remain on the differential diagnosis list. Brain tumours are often classified according to the cell lineage they originate from (meninges, glial cells, choroid plexus cells, lymphocytes, ependymal cells, etc.) or based on the circumferential location of the tumour in relation to the brain (extra-axial, intra-axial or intraventricular). In both dogs and cats, the most common intracranial tumours are meningiomas and typically present as single extra-axial masses, while the most common intra-axial tumours are gliomas (astrocytomas, oligodendrogliomas, glioblastoma). Round cell tumours such as lymphoma or histiocytic sarcoma can present with an intra-axial, intraventricular or extra-axial localisation (Dickinson 2014). The mechanisms of injury to the brain are multiple and depend on the size, growth and location of the tumour: direct invasion of the brain tissue and loss of function, ischaemia, focal compression, oedema formation and increased intracranial pressure, alteration of the balance of neurotransmitters, but also biochemical and immunological derangements. The strong breed predisposition seen for some tumours as gliomas in some dog breeds such as boxers, bulldogs and Boston terriers suggests a genetic susceptibility in specific breeds (Truvé et al. 2016).

Clinical Signs The clinical signs depend not only on the location of the neoplasm but also on the rate of its growth and the local biochemical changes as mentioned above. It is not uncommon to diagnose a sizeable mass in the light of relatively subtle signs, as the brain is able to compensate for substantial changes in pressure and volume before the compensation mechanisms are exhausted. Behavioural changes are common and do not infrequently precede the development of neurological deficits that will make a forebrain neurolocalisation more obvious. As an example, the most common clinical signs in a study of pituitary macroadenoma were mentation (depression, stupor) and behavioural changes (pacing, disorientation, circling, head pressing, aggression). Behavioural changes can also be the only clinical complaint. Masses affecting the limbic system, association areas (such as the frontal lobe and parieto-temporal junction) or the ascending reticular activating system are likely to produce behavioural changes with or without any obvious neurological deficits (Berns et al. 2015). The changes that can be seen include the development of fearful or aggressive behaviour or conversely the development of an attention-seeking behaviour in a previously fearful animal. Cats can be seen hiding more in unusual places or can be found stuck in a corner of a room. Head pressing or compulsively circling to one side only will raise the suspicion towards a significant intracranial lesion, with the former signs being suggestive of raised intracranial pressure. House soiling can also precede the development of more worrisome neurological signs and could be exacerbated by the presence of a concurrent endocrinopathy causing PU/PD (hyperadrenocorticism, central diabetes insipidus or acromegaly due to a secreting pituitary tumour). Raised intracranial pressure can also lead to clinical signs of pain including vocalisation, aggression, lethargy and head pressing behaviours (Madison et al. 2015).

A history of seizure is a specific sign for a forebrain lesion and can be seen in the absence of neurological deficits despite the presence of a substantial lesion. Neurological deficits will vary depending on the location and size of the mass and can suggest a focal or multifocal pathology. Due to the crossover of the sensory and motor pathways, a lateralised forebrain lesion will manifest as neurological deficits on the contralateral side of the body (absent vision, abnormal proprioception or nasal sensation).

Clinical signs that overlap with CDS:

- Depression
- Pacing
- House soiling
- Altered interactions

Clinical signs that would not overlap with CDS:

- Neurological deficits
- Seizures
- Signs of raised intracranial pressure

Diagnosis The cornerstone of the diagnosis of intracranial neoplasia remains advanced imaging (with MRI being preferred to CT). This is performed after having ruled out extracranial causes of forebrain or brainstem dysfunction. A presumptive diagnosis of a brain tumour is often reached after considering the history, signalment and imaging findings along with the cerebrospinal fluid results. Occasionally, repeated imaging or brain biopsies may be necessary to reach a more definitive diagnosis or to rule out other non-neoplastic neoplastic differentials (Fig. 2.2).

Treatment Treatment options include surgery, chemotherapy, radiation therapy, palliative therapy or a combination of these depending on the clinical presentation, location, size and nature of the tumour. While the prognosis is guarded at best, remarkable results can be achieved in selected patients following surgery (extra-axial meningiomas in cats, pituitary tumours) or following a combination of surgery and radiation therapy (extra-axial meningiomas in dogs) or with chemotherapy (lymphoma) (Madison et al. 2015).

2.2.4.9 Inflammatory Non-infectious Brain Disease

Aetiopathogenesis The pathophysiology of non-infectious inflammatory brain disease is still poorly understood. An immune-mediated mechanism following an overreaction of the immune system due to antigen mimicry is often postulated, but a multifactorial process is likely. The fact that a strong breed predisposition is present in dogs makes a genetic basis a very popular theory (supported by the identification of a gene locus associated with an increased risk of developing necrotising meningoencephalitis in pug dogs). Non-infectious inflammatory brain diseases are less common in cats but are likely under-reported or underdiagnosed (Zeira et al. 2015).

The non-infectious meningoencephalitis is often described under the umbrella "meningoencephalitis of unknown aetiology/origin = MUE/MUO". This covers a number of pathological lesions, for example, necrotising meningoencephalitis, necrotising leucoencephalitis, eosinophilic meningoencephalitis and granulomatous meningoencephalitis.

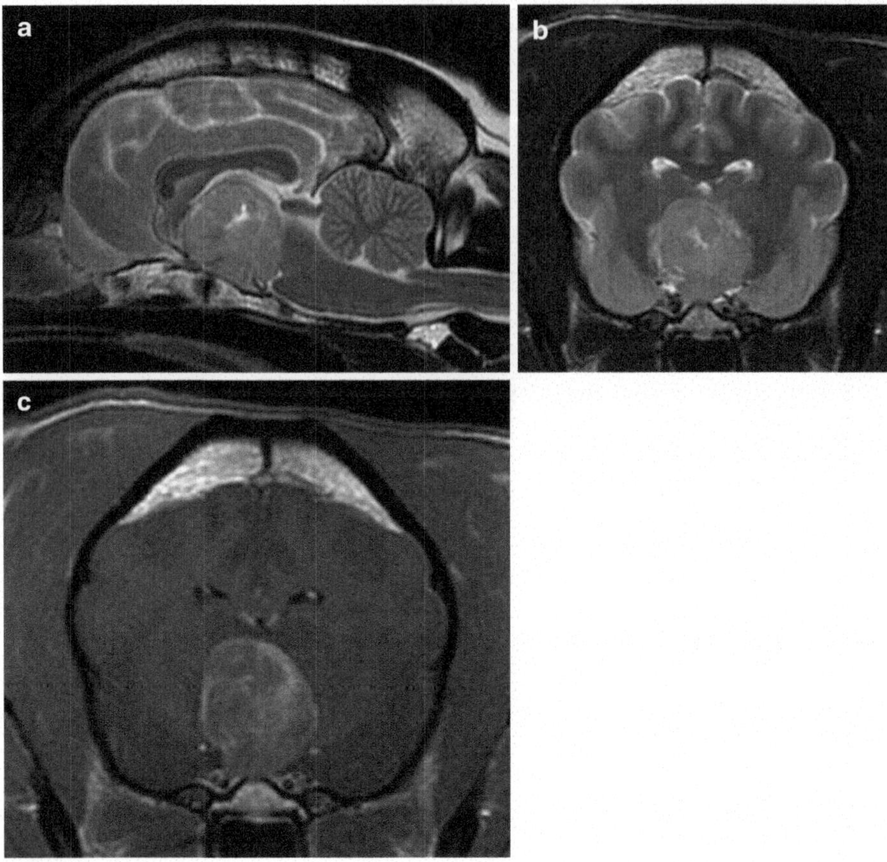

Fig. 2.2 Sagittal T2 (**a**), transverse T2 (**b**) and transverse T1 post-contrast (**c**) images of a 9.5-year-old border terrier presenting with progressive behavioural changes alone showing a large mass in the pituitary region

Clinical Signs The clinical signs again reflect the anatomical location of the lesion(s) and are often a combination of neurological signs (neurological deficits, seizures, signs of raised intracranial pressure) and behavioural changes. While typically the presentation involves a middle-aged patient with clinical signs that can progress rapidly, some animals will present later in age with insidious and non-specific clinical signs as being less responsive, pacing, lethargic or unwilling to move. Many of the inflammatory diseases seem to have a predisposition for the brainstem and cerebellum, so cerebello-vestibular signs are relatively common with this diagnosis (Han et al. 2015).

Clinical signs that overlap with CDS:

- Lethargy
- Pacing
- Loss of trained behaviours

Clinical signs that would not overlap with CDS:

- Neurological deficits
- Seizures
- Signs of raised intracranial pressure

Diagnosis/Exclusion The diagnosis is based on exclusion of extracranial causes of forebrain or brainstem dysfunction supported by advanced imaging and CSF analysis and after ruling out the common infectious pathogens present in the geographical area. Ante-mortem diagnosis is often speculative after exclusion of infectious causes. However, the advances of stereotactic brain biopsy techniques have recently permitted the gain of more definitive ante-mortem diagnosis in carefully selected patients.

Treatment The treatment of MUE/MUO is centred on immunosuppression with prednisolone as the first-line drug. The use of adjunctive immunosuppressives and which drug to use is a more debated topic. Following treatment, the clinician is often presented with three case scenarios: one population of patients will be well controlled and weaned off medications over the course of 3–6 months, one population of animals will require lifelong or adjunctive immunosuppression, and one population may fail and deteriorate despite treatment.

2.2.4.10 Infectious Brain Disease

Aetiopathogenesis Infectious agents involved can be bacterial, viral, protozoal or even fungal or algal depending on the geographical region involved. The route of entry of the pathogen can be by local extension from a nearby infection (ocular, nasal, penetrating wound) or could be secondary to systemic circulation of the agents and penetration through the blood CSF or blood-brain barrier (Ebani et al. 2014).

When a pathogen is present, the tissue injury is not only the consequence of the pathogen's direct insult but almost more importantly reflects a detrimental response of the immune system to this pathogen. The inflammatory reaction may lead to the creation of abscess or empyema (bacterial diseases), granulomas (*Toxoplasma* infection, fungal diseases, mycobacterial infection), brain atrophy or necrosis (*Neospora* infection), vasculitis (FIP virus) or demyelination (distemper) (Ebani et al. 2014).

While young unvaccinated animals are more at risk, cats or dogs of any age can be affected, and a latent infection can reactivate later in life (e.g. old dog distemper).

Pathogens involved in infectious meningoencephalitis in dogs and cats.

Cats:

- Bacterial: rods, cocci, mycobacteria
- Viral: FeLV, FIV, FIP, bornavirus
- Protozoal: *Toxoplasma gondii*
- Fungal: *Blastomyces*, *Cryptococcus*, *Coccidioidomycosis*
- Parasitic: *Taenia serialis*, *Cuterebra*, *Dirofilaria immitis*

Dogs:

- Bacterial: rods, cocci, mycobacteria, rickettsia
- Viral: distemper
- Protozoal: *Toxoplasma gondii*, *Neospora caninum*, *Leishmania*, *Toxoplasma gondii*
- Fungal: *Aspergillus*, *Cryptococcus*, *Blastomyces*, *Coccidioidomycosis*
- Parasitic: *Angiostrongylus vasorum*, *Baylisascaris procyonis*, *Dirofilaria immitis*

Clinical Signs The clinical signs are variable and reflect the anatomical location of the lesion(s) present. The neurolocalisation can indeed reflect focal or multifocal pathology. Neurological deficits are commonly noticed but could be mild and can be overshadowed by more conspicuous neurological signs such as seizures or behavioural changes. Behavioural changes can manifest as an altered mental status, reduced activity or lethargy, especially if the ascending activating reticular formation is involved either in the brainstem or in both cerebral hemispheres. Circling, pacing, head pressing and excessive panting or restlessness are also not uncommon.

Clinical signs that overlap with CDS:

- Altered mental status
- Reduced activity or lethargy

Clinical signs that would not overlap with CDS:

- Neurological deficits
- Seizures
- Signs of raised intracranial pressure

Diagnosis/Exclusion The diagnosis is reached after exclusion of metabolic causes of forebrain and/or forebrain dysfunction and ideally following advanced imaging. MRI is the modality of choice and should ideally be performed before spinal fluid collection or analysis is performed. Blood serology, CSF culture and PCRs for specific agents can be performed to try to identify the causative pathogen. This should take into consideration the geographical area of residence of the pet but also the relevant travel history.

Treatment The treatment is based on the pathological agent identified and the presence or not of a discrete lesion. Surgery can be indicated in the presence of a focal abscess or granuloma for both diagnostic and therapeutic purposes. A specific medical therapeutic plan can be instaured once the pathogens have been identified and, ideally, when culture and sensitivity results are available.

2.2.4.11 Traumatic Differential: Chronic Repetitive Traumatic Brain Injury

Aetiopathogenesis Chronic repetitive traumatic brain injury occurs in professional sports such as American football or boxing (called chronic traumatic encephalopathy) but remains exceedingly rare in small animal patients when it almost invariably results from animal abuse. The brain damage usually occurs following the development of subarachnoid haemorrhage, lacerations of the brain tissue and secondary vasogenic oedema.

Clinical Signs While only one case report described the clinical signs of a dog with repetitive brain trauma due to physical abuse, people may show mainly behavioural signs, potentially years after the injuries stopped, with signs of progressive dementia and depression. The dog in the published report showed signs of decreased mentation, disorientation, absence of response to sound and acoustic stimuli, generalised proprioceptive ataxia, neurological deficits pointing towards a forebrain neurolocalisation and tremors (Plessas et al. 2013).

Clinical signs that overlap with CDS:

- Behavioural changes
- Disorientation
- Decreased mentation

Clinical signs that would not overlap with CDS:

- Tremors
- Generalised proprioceptive ataxia
- Neurological deficits

Diagnosis/Exclusion The diagnosis is supported by the history, physical examination with evidence of repetitive trauma, and evidence of brain atrophy on CT or MRI. Spinal fluid results may reveal a neutrophilic pleocytosis.

Treatment There is no specific treatment.

2.2.4.12 Toxic Differential: Lead Poisoning

Aetiopathogenesis The heavy metal causes direct damage to blood vessels leading to haemorrhage ischaemia (with subsequent necrosis) and oedema of the central nervous system. It is also postulated that lead or other secondary metabolic substances may cause neuronal damage (Pauli and Buskirk 2007).

Clinical Signs Dogs usually present with GI signs (vomiting and diarrhoea), and these are usually seen in combination with CNS disease, but rarely the CNS signs will be the only ones present. In this case seizures are common, but behavioural/mentation changes with hysteria can occur in isolation alongside other signs of tremors, ataxia, hypersensitivity, champing of the jaws and tics.

Clinical signs that overlap with CDS:

- Mentation/behaviour changes with hysteria

Clinical signs that would not overlap with CDS:

- Seizures
- GI signs
- Tremors
- Ataxia
- Champing of the jaws and tics (which in the author's opinion are interpreted as a myoclonus)

Diagnosis/Exclusion Appropriate clinical signs alongside increased levels of lead in the blood.

Treatment Chelation therapy using calcium EDTA or penicillamine and antiepileptics where seizures are present is preferred.

2.2.4.13 Vascular

Cerebrovascular disease defines any brain pathology that results from pathology of its blood supply. The major pathological processes include blood vessel disruption (leading to haemorrhage), blood vessel obstruction (leading to ischaemia or infarction), blood vessel malformation and vasculitis.

Hypertensive Encephalopathy

Aetiopathogenesis Hypertensive encephalopathy is due to blood vessel damage following sustained or acute periods of hypertension. The blood vessels under stress may narrow or become harder (atherosclerosis) or thinner, leading to blood vessel obstruction, rupture or leakage. This can ultimately lead to the development of haemorrhagic or ischaemic strokes which are peracute manifestations or neurological signs following damage to the brain blood vessels. Other proposed mechanisms of injury include the failure of the normal autoregulation of the vasculature leading to hyperperfusion and secondary vasogenic oedema. Target organ damage is possible when the systolic blood pressure exceeds 150–160 mmHg, but some authors suggest that a systolic blood pressure over 180 mmHg may be more relevant to the brain (Panciera 2000).

Clinical Signs The clinical signs can initially be vague. Animals can present for non-specific lethargy, altered mentation, decreased vision or decreased ability to find their food. Neurological deficits (localising the problem to the forebrain and or brainstem) may ensue, with some animals developing signs suggestive of raised intracranial pressure (head pressing, decerebrate posture, opisthotonus). Retinal haemorrhages, tortuous blood vessels, multifocal retinal oedema or retinal detachment may be detected on ophthalmological examination of the fundus. Some patients may present with peracute episodes against a background of more insidious illness. Ischaemic or haemorrhagic strokes can also cause an acute onset of neurological signs that may be witnessed or missed by the owners. Neurological signs such as circling, pacing or decreased vision could be residual signs perceived by the owners as a decline in cognitive function. Most clinical signs attributable to a stroke would be expected to improve with time, although recurrent or multiple cerebrovascular accidents (CVAs) are possible and can complicate the clinical picture (Cain and Khalil 2002).

Clinical signs that overlap with CDS:

- Lethargy
- Altered mentation
- Inability to find food

Clinical signs that would not overlap with CDS:

- Neurological deficits
- Signs of raised intracranial pressure

Diagnosis/Exclusion The diagnosis of hypertensive encephalopathy is based on the identification of a systolic blood pressure over 160–180 mmHg in the presence of neurological deficits suggestive of brain dysfunction together with other clinical signs (hypertensive retinopathy) or imaging findings (white matter oedema, ischaemic or haemorrhagic stroke) suggestive of hypertension in the absence of other pathological processes. After a clinical suspicion of hypertensive encephalopathy due to systemic hypertension is established, the clinician should focus on the diagnosis of the underlying primary pathological process. Indeed, in dogs and cats, systemic hypertension is seldom idiopathic (although possibly seen in up to 20% of cats) and often reflects endocrinopathy (hyperthyroidism, hyperadrenocorticism, diabetes mellitus, hyperaldosteronism, acromegaly, phaeochromocytoma), cardiomyopathy (hypertrophic cardiomyopathy in cats), renal insufficiency or obesity (Stepien et al. 2003).

Treatment The treatment of hypertensive encephalopathy should be focused on the treatment of the underlying primary cause. While angiotensin-converting enzyme (ACE) inhibitor forms the first-line treatment of hypertension in dogs, calcium channel blockers as amlodipine are often necessary to bring the blood pressure within physiological ranges in cats.

Cerebrovascular Accidents

Aetiopathogenesis As mentioned above, CVAs are the peracute clinical manifestation of neurological signs due to a non-traumatic blood vessel rupture (haemorrhagic stroke) or blood vessel occlusion (ischaemic stroke). Causes of blood vessel disruption include coagulopathies, hypertension, neoplasia, and inflammatory or infectious processes. Blood vessels can get occluded by a thrombus formed in situ or at a distant site (thromboembolism), a neoplastic or septic emboli or fat embolisation. Some of the common diseases implicated in the pathogenesis of ischaemic or haemorrhagic strokes in dogs and cats are listed below. However, in a significant number of cases, no underlying cause can be identified (Jensen et al. 1997).

Causes of ischaemic stroke in dogs and cats:

- Chronic kidney disease
- Hypertension
- Diabetes mellitus
- Neoplasia
- Protein-losing nephropathy or enteropathy
- Cushing's disease
- Cardiomyopathy
- Hypothyroidism (dogs)

Common causes of haemorrhagic stroke in dogs and cats:

- Angiostrongylus vasorum infection (dogs)
- Neoplasia
- Hypertension
- Coagulopathy
- Bacterial infections
- Brain atrophy

Clinical Signs The clinical signs will reflect the localisation of the lesion(s) present. The neurolocalisation can therefore be focal or multifocal. The thalamus and cerebellum are two regions of predilection for ischaemic stroke. Clinical signs present can include an altered mentation, pacing, compulsive circling, vocalising, hemineglecting or bumping into things. The onset tends to be peracute with a progressive improvement of the signs. However, in the case of multiple CVAs, the signs can be persistent waxing and waning. Moreover, the onset can be missed and the neurological deficits can persist, leaving the pet debilitated. The owner may bring their pet to the clinician with the complaint of an altered mentation or recent-onset dementia.

Diagnosis/Exclusion The diagnosis is based on the identification of ischaemic or haemorrhagic lesion(s) on MRI. CT can be valuable in the identification of intracranial haemorrhage, but MRI will be superior as it allows the identification of neoplastic processes that can cause secondary intracranial haemorrhage (e.g. some gliomas or meningiomas). The identification of a stroke warrants further investigation to rule out an underlying cause. This may involve blood pressure monitoring, additional blood testing, endocrine testing, urinalysis (including a Urine Protein Creatinine Ratio (UPC) and imaging of the thorax and abdomen. In a substantial number of cases, no underlying cause can be identified. This subset of patients seems to be at a decreased risk for suffering further episodes in the future (Kraft and Egner 2003).

Treatment The treatment should be aimed at the underlying pathological process present and is otherwise supportive. Most of the current research is trying to identify treatment options to reduce the secondary injury occurring in the brain, but to date no effective treatment is available (Jensen 1997; Kraft and Egner 2003).

2.2.5 Diagnostic Testing

2.2.5.1 Laboratory Testing

The following step is the investigation of possible underlying metabolic conditions (Table 2.7). Most metabolic conditions will cause symmetrical neurological deficits if any, but this is not a hard and fast rule. Asymmetrical signs, including circling or vestibular signs (including positional nystagmus), can be seen with metabolic diseases as hepatic encephalopathy (although vestibular signs alone are not usually expected with hepatic encephalopathy). The minimum initial database should include a blood haematology with differential and cytology to investigate the presence of anaemia (congenital portosystemic shunt, hypothyroidism, anaemia of chronic disease, lead poisoning, etc.), an abnormal white blood cell count (inflammatory or neoplastic process) or abnormal platelet count (e.g. paraneoplastic thrombocytopenia, thrombocytosis associated with hyperadrenocorticism) (Ettinger and Feldman 2009).

A serum biochemistry is recommended to evaluate renal (urea, creatinine) and liver (ALT, ALP) parameters but should also include blood electrolytes, serum albumin, globulin and serum cholesterol. Repeated fasting blood glucose measurements are performed (at least twice in a 24–48 h period) after at least 12 h of fasting in animals presenting with intermittent or persistent alteration of consciousness (especially if the neurological examination is normal) to rule out hypoglycaemia. Serum ammonia and pre- and post-prandial bile acid are performed in animals with raised liver enzymes or in the presence of forebrain or brainstem signs or altered mental

Table 2.7 Metabolic conditions to rule out

Condition	Common causes	Testing suggested
Hypoglycaemia	Insulinoma (dogs), neoplasia, Addison	Repeated fasting blood glucose, fructosamine
Hyperglycaemia	Diabetes mellitus, hyperadrenocorticism	Blood glucose
Hepatic encephalopathy	Congenital portosystemic shunt, chronic hepatopathies and acquired portosystemic shunt	Fasting blood ammonia, dynamic bile acid stimulation test
Hyperthyroidism	Thyroid adenoma (cats)	Total T4
Uremic encephalopathy	Renal insufficiency	Serum urea, creatinine, SDMA, USG
Hypothyroidism	Lymphoplasmacytic adenitis, idiopathic atrophy (dogs)	Total T4, TSH
Hypernatraemia	Hyperaldosteronism (cats), adipsia, diabetes insipidus	Serum sodium
Hyponatraemia	Hyperadrenocorticism, GI or renal losses, cardiac or hepatic insufficiency	Serum sodium
Hypercalcaemia	Paraneoplastic, osteolytic lesions, hyperparathyroidism, renal insufficiency	Ionised calcium
Hypocalcaemia	Hypoparathyroidism, nutritional secondary hyperparathyroidism, renal insufficiency	Ionised calcium

status (hyper or hypo). Serum thyroid level (total T4) is assessed in patients with altered mental statuses and accompanying relevant clinical signs to investigate the presence of hypothyroidism (dogs) or hyperthyroidism (cats). It is worth noting that clinical signs of forebrain disease are only reported in dogs with myxoedema coma and multiple vascular events.

Any animal with polyuria, polydipsia, periuria, dysuria or house soiling with inappropriate urination should have a complete urinalysis with urine cytology, UPC and urine culture performed to document the presence of crystalluria, urinary tract infection or proteinuria. Animals with gastrointestinal signs or passing faeces in the house should have a rectal examination performed as well as a faecal analysis.

2.2.5.2 Blood Pressure

Non-invasive blood pressure is recommended in older animals presenting with vague, non-specific or waxing and waning signs of altered consciousness or behaviour. This is often repeated in a quiet environment to increase the accuracy of the readings. Consistently elevated blood pressure should warrant further investigation to rule out the underlying cause, provided the readings are not due to the so-called white coat effect. If the latter is suspected, the measurements should be performed if possible at home while the animal is most relaxed (Nelson and Couto 2014).

Causes of hypertension in older animals:

- Chronic renal insufficiency
- Hyperthyroidism (cats)
- Hyperadrenocorticism (dogs)
- Diabetes mellitus
- Cardiomyopathy
- Phaeochromocytoma
- Hyperaldosteronism (cats)
- Idiopathic
- Raised intracranial pressure

2.2.5.3 Advanced Imaging

Whether or not the neurological examination is normal, in the presence of behavioural signs, an MRI scan of the brain should be considered before establishing a presumptive diagnosis of CDS. Advanced imaging is performed after having ruled out the potential metabolic causes mentioned in the previous two paragraphs. MRI is the modality of choice to assess the brain and will be superior to CT due to its better soft tissue resolution. The results of the MRI scan need to be interpreted in conjunction with the index of suspicion and additional tests. If the MRI scan is normal, then CSF is collected. Patients affected by CDS can have a normal brain MRI or evidence of brain atrophy and ventricular enlargement. However, MRI changes of brain atrophy are not pathognomonic and can be seen with some chronic metabolic (chronic hypoglycaemia, congenital PSS), inflammatory (eosinophilic meningoencephalitis), neurodegenerative (e.g. ceroid lipofuscinosis) or even infectious diseases (*Neospora caninum* infection). Consequently, MRI findings of brain atrophy are not diagnostic of CDS but are compatible with it (Ettinger and Feldman 2009).

Differential diagnosis for brain atrophy on MRI includes CDS; degenerative storage diseases, in particular ceroid lipofuscinosis; congenital portosystemic shunt; chronic or repeated metabolic/toxic disturbances, for example, chronic hypoglycaemia; eosinophilic meningoencephalitis; *Neospora* infection; and chronic repetitive traumatic brain injury (Fig. 2.3).

MRI changes compatible with CDS:

- Normal MRI
- Ventricular system dilation
- Cortical atrophy—prominent sulci
- Narrowing of the interthalamic adhesion

2.2.5.4 CSF Analysis

The next step consists in performing a CSF analysis to rule out neoplastic or inflammatory conditions. If judged safe, the CSF analysis should be performed irrespective of the MRI findings as an inflammatory or neoplastic process can be present despite a normal MRI scan. An elevated cell count or protein level will be suggestive of an inflammatory process, although vascular, neoplastic or degenerative diseases can also cause such elevations. Cytological analysis may allow the identification of infectious agents, abnormal inclusions (e.g. viral) or neoplastic cells (e.g. lymphoma) (Nelson and Couto 2014).

2.2.5.5 Additional Testing

If the neurological examination, MRI scan and CSF analysis are normal, then an underlying pathological process such as inflammation, infection or neurodegenerative disease will be considered very unlikely. A presumptive diagnosis of CDS could be established at this stage. If in the neurological examination the MRI scan and/or the CSF results are abnormal, then an underlying pathological process other than CDS is suspected, and further testing is warranted.

In the presence of brain atrophy alone (despite a normal CSF analysis), the clinician should focus again on ruling out chronic hypoglycaemia by repeating dome fasting blood glucose sampling, being sure a dynamic bile acid stimulation test was correctly performed and was normal, and by ruling out neurodegenerative causes such as ceroid lipofuscinosis (Ettinger and Feldman 2009).

Blood serology against infectious agents or PCRs looking for DNA for such agents (e.g. feline coronavirus, canine distemper virus, *Neospora caninum*, etc.) are performed when the MRI scan or spinal fluid results suggest that an inflammatory process is present. PARR testing (PCR for lymphocyte antigen receptor rearrangement) or flow cytometry can also be valuable for the diagnosis of lymphoma in the presence of an atypical population of lymphocytes, while genetic testing for a known mutation associated with degenerative diseases (e.g. identification of one of the known breed-specific mutations identified for ceroid lipofuscinosis or L2-HGA) may offer a definitive diagnosis (Nelson and Couto 2014).

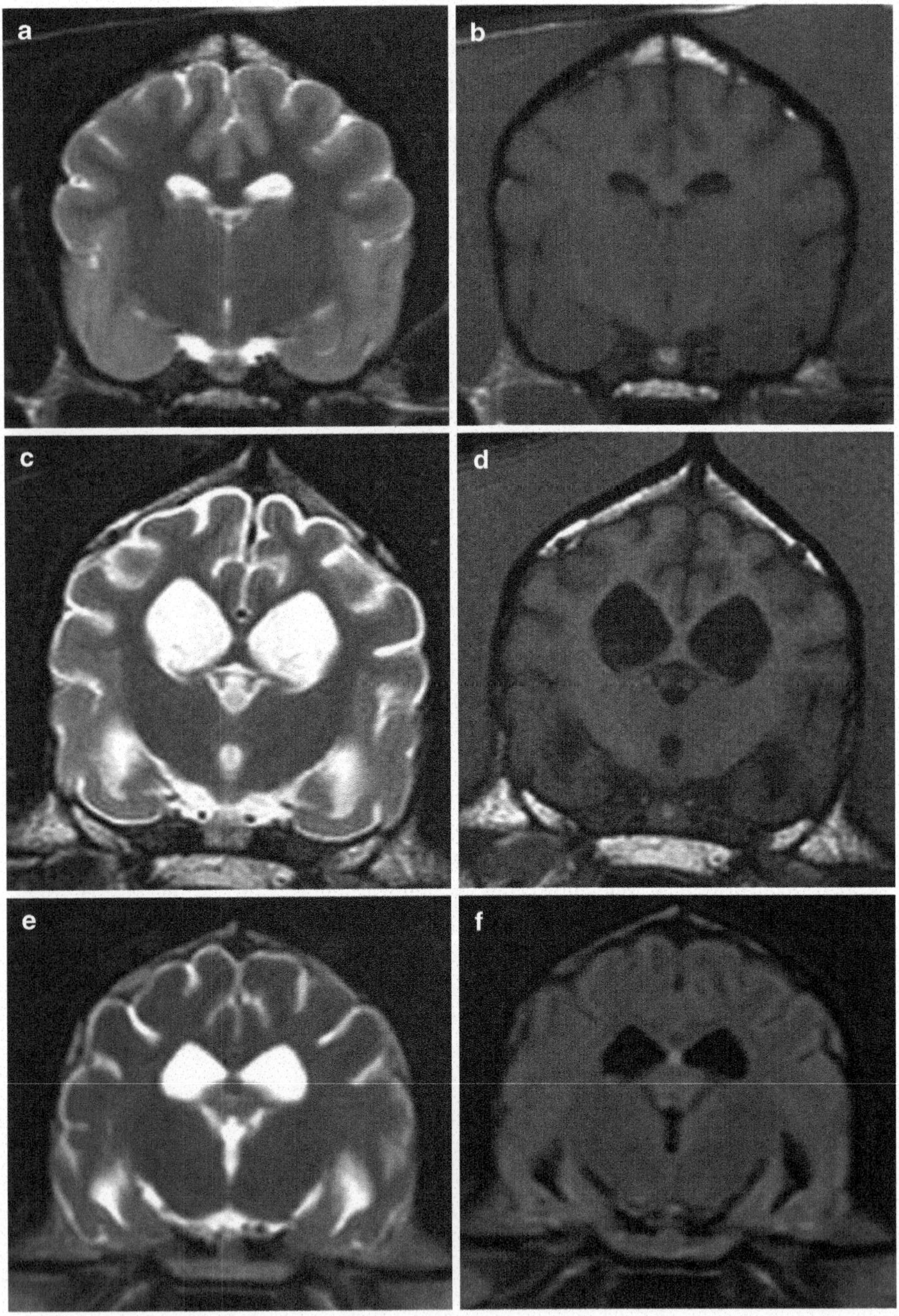

Fig. 2.3 Transverse T2 (**a**)- and T1 (**b**)-weighted images of a normal dog. Transverse T2 (**c**)- and T1 (**d**)-weighted images of a dog suffering from ceroid lipofuscinosis. Transverse T2 (**e**)- and FLAIR (**f**)-weighted images of a dog affected by CDS. All sections are at the level of the interthalamic adhesion

Urine screening for metabolic diseases will allow the identification of organic acidurias (e.g. L2-HGA) and some other storage diseases, as will biopsy of other organs (e.g. liver, skin, etc., depending on the storage disease).

2.2.6 Summary

The following chart summarises the clinical approach taken to rule out the differential diagnosis for CDS in dogs and cats. As mentioned above, a strong suspicion for CDS exists when all the diagnostic testing rules out other differentials (Nelson and Couto 2014).

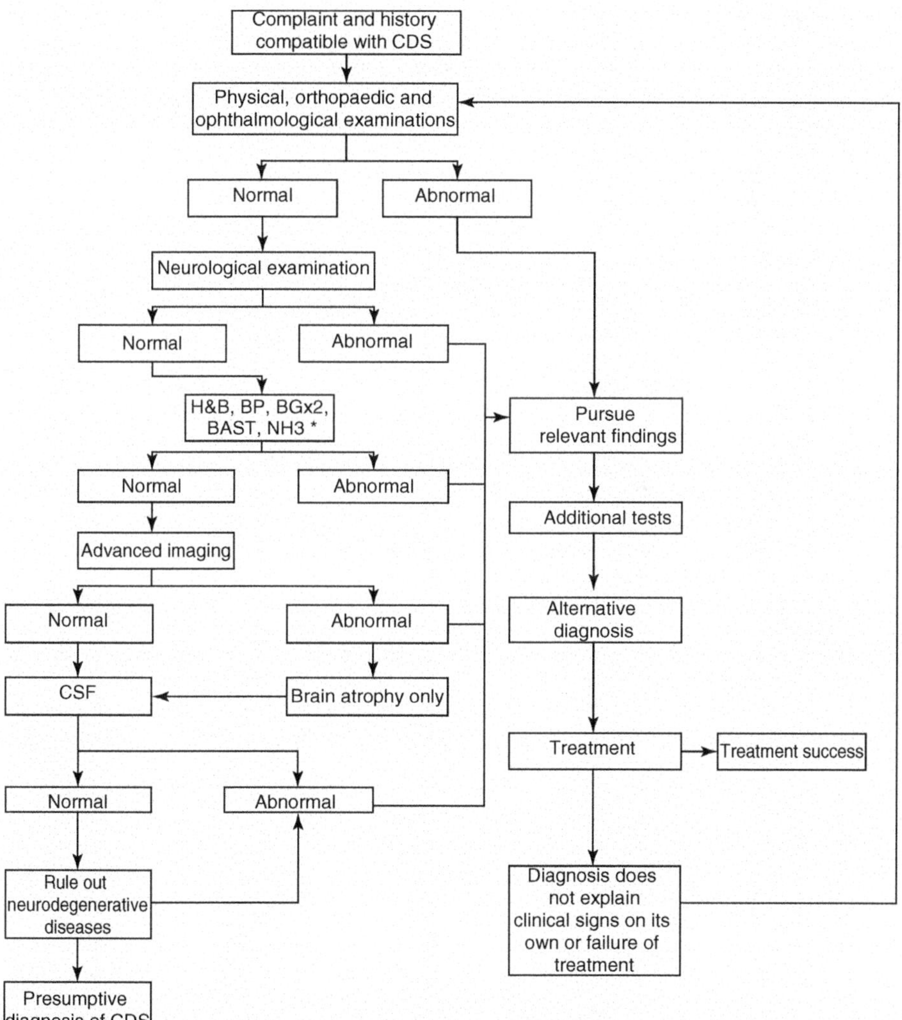

* *H&B* Hematology and Biochemisty, *BP* Blood pressure, *BGx2* blood glucose X 2, *BAST* Bile acid stimulation test, *NH3* Ammonia

References

Attems J (2005) Sporadic cerebral amyloid angiopathy: pathology, clinical implications, and possible pathomechanisms. Acta Neuropathol 110(4):345–359

Bain M (2016) Housesoiling - Dogs. In: Tilley LP and Smith FWK (eds) Blackwell's Five-Minute Veterinary Consult: Canine and Feline 6th ed. Wiley & Sons, Ames IA,

Bécuwe-Bonnet V, Bélanger M-C, Diane Frank, Parent J, Hélie P (2012) Gastrointestinal disorders in dogs with excessive licking of surfaces. Journal of Veterinary Behavior: Clinical Applications and Research 7 (4):194–204

Berns GS, Brooks AM, Spivak M (2015) Scent of the familiar: An fMRI study of canine brain responses to familiar and unfamiliar human and dog odors. Behav Process 110:37–46

Bowen J, Heath S (2005) Behaviour Problems in Small Animals: Practical Advice for the Veterinary Team Elsevier Saunders, Edinburgh

Cain AE, Khalil RA (2002) Pathophysiology of essential hypertension: role of the pump, the vessel and the kidney. Semin Nephrol 22(1):3–16

Camps T, Amat M, Mariotti VM, Brech SL, Manteca X (2012) Pain-related aggression in dogs: 12 clinical cases. Journal of Veterinary Behavior: Clinical Applications and Research 7(2):99–102

Denenberg S, Landsberg GM (2016) Excessive vocalization and waking at nights. In: Tilley LP and Smith FWK (eds) Blackwell's Five-Minute Veterinary Consult: Canine and Feline 6th ed. Wiley & Sons, Ames IA, pp 470–471

Dickinson PJ (2014) Advances in diagnostic and treatment modalities for intracranial tumors. J Veterinary Int Med 28(4):1165

Ebani VV, Bertelloni F, Torracca B, Cerri D (2014) Serological survey of Borrelia burgdorferi sensu lato, Anaplasma phagocytophilum, and Ehrlichia canis infections in rural and urban dogs in Central Italy. Ann Agric Environ Med 21(4):671–675

Ettinger SJ, Feldman EC (2009) Textbook of veterinary internal medicine, 7 edn. ISBN 13: 978-1-4160-6593-7

Farmer-Dougan V, Quick A, Harper K, Schmidt K, Campbell D (2014) Behavior of hearing or vision impaired and normal hearing and vision dogs (Canis lupis familiaris): Not the same, but not that different. Journal of Veterinary Behavior: Clinical Applications and Research 9(6):316–323

Frank D (2014) Recognizing Behavioral Signs of Pain and Disease. Veterinary Clinics of North America: Small Animal Practice 44(3):507–524

Gold AJ, Scott MA, Fyfe JC (2015) Failure to thrive and life-threatening complications due to inherited selective cobalamin malabsorption effectively managed in a juvenile Australian shepherd dog. Canad Veterinary J 56(10):1029–1034

Gould E (1999) Serotonin and Hippocampal Neurogenesis. Neuropsychopharmacology 21 (2):46S-51S

Gruen M, Sherman BS (2016) Housesoiling - Cats. In: Tilley LP and Smith FWK (eds) Blackwell's Five-Minute Veterinary Consult: Canine and Feline 6th ed. Wiley & Sons, Ames IA

Han J, Chang D, Na K (2015) A multiplex quantitative real-time polymerase chain reaction panel for detecting neurologic pathogens in dogs with meningoencephalitis. J Veterinary Sci 16(3):341–347

Head E, Nukala VN, Fenoglio KA, Muggenburg BA, Cotman CW, Sullivan PG (2009a) Effects of age, dietary, and behavioral enrichment on brain mitochondria in a canine model of human aging. Exp Neurology 220(1):171–176

Head E, Nukala VN, Fenoglio KA, Muggenburg BA, Cotman CW, Sullivan PG (2009b) Effects of age, dietary, and behavioral enrichment on brain mitochondria in a canine model of human aging. Exp Neurol 220(1):171–176

Holt DE, Washabau RJ, Djali S, Dayrell-Hart B, Drobatz KJ, Heyes MP, Robinson MB (2002) Cerebrospinal fluid glutamine, tryptophan, and tryptophan metabolite concentrations in dogs with portosystemic shunts. Am J Veterinary Res 63(8):1167–1171

Jensen JL, Henik RA, Brownfield M, Armstrong J (1997) Plasma rennin activity and angiotensin I and aldosterone concentrations in cats with hypertension associated with chronic renal failure. Am J Veterinary Res 58(5):535–541

Johnson LN, Heinze CR, Linder DE, Freeman LM (2015) Evaluation of marketing claims, ingredients, and nutrient profiles of over-the-counter diets marketed for skin and coat health of dogs. J Am Veterinary Med Assoc 246(12):1334–1338

Kraft W, Egner B (2003) Causes and effects of hypertension. In: Egner B, Carr A, Brown S (eds) Facts of blood pressure in dogs and cats. Beate Egner Vet Verlag, Babenhausen, pp 61–86

Landsberg G (2005) Therapeutic agents for the treatment of cognitive dysfunction syndrome in senior dogs. Progress in Neuro- Psychopharmacology and Biological Psychiatry 29(3):471–479

Landsberg G, Araujo JA (2005) Behavior problems in geriatric pets. Veterinary Clin N Am-Small Animal Pract 35(3):675–698

Landsberg G, Denenberg S (2009) Behaviour problems in the senior pet. In: Horwitz D, Mills DS (eds) BSAVA Manual of Canine and Feline Behavioural Medicine, 2nd edn. BSAVA, Gloucester, pp 127–135

Landsberg GM, Denenberg S, Araujo JA (2010) Cognitive dysfunction in cats: a syndrome we used to dismiss as 'old age'. J Feline Med Surg 12(11):837–848

Landsberg GM, DePorter T, Araujo JA (2011) Clinical signs and management of anxiety, sleeplessness, and cognitive dysfunction in the senior pet. Veterinary Clin N Am: Small Animal Pract 41(3):565–590

Landsberg GM, Nichol J, Araujo JA (2012) Cognitive dysfunction syndrome a disease of canine and feline brain aging. Veterinary Clin N Am-Small Animal Pract 42(4):749–768

Landsberg GM, Hunthausen W, Ackerman LA (2013) The effect of aging on behavior in senior pets. In: Behavior problems of the dog and cat, 3rd edn. Saunders, Edinburgh, pp 211–235

Luescher AU (2003) Diagnosis and management of compulsive disorders in dogs and cats. Veterinary Clinics of North America: Small Animal Practice 33(2):253–267

Madison AM, Sharma A, Haidekker MA (2015) Noninvasive assessment of intracranial pressure in dogs by use of biomechanical response behavior, diagnostic imaging, and finite element analysis. Am J Veterinary Res 76(8):667–678

Manteca X (2011) Nutrition and Behavior in Senior Dogs. Topics in Companion Animal Medicine 26(1):33–36

Meleo KA, Peterson ME (2014) Treatment of Insulinoma in the Dog, Cat, and Ferret. In: Bonagura JD, Twedt DC (eds) Current Veterinary Therapy XV, Chapter: Web Chapter 23 (www.currentveterinarytherapy.com), Saunders Elsevier, pp e130–e134

Neilson JC (2004) Feline house soiling: Elimination and marking behaviors. Clinical Techniques in Small Animal Practice 19(4):216–224

Nelson RW, Couto CG (2014). Small animal internal medicine, 5th edition. ISBN 978-0-323-08682-0

Norton RD, Lenox CE, Manino P, Vulgamott JC (2016) Nutritional considerations for dogs and cats with liver disease. J Am Animal Hosp Assoc 52(1):1–7

Ohl F, Arndt SS, van der Staay FJ, (2008) Pathological anxiety in animals. The Veterinary Journal 175 (1):18–26

Overall KL (2005) Mental Illness in animals. In: McMillan F (ed) Well-being animals. ISU Press, Ames IA, pp 127-143

Panciera DL (2000) Cardiovascular complications of thyroid disease. In: Bonagura JD (ed) Kirk's current veterinary therapy XIII, small animal practice, 13th edn. WB Saunders, Philadelphia, pp 716–719

Pauli JN, Buskirk SW (2007) Recreational shooting of prairie dogs: a portal for lead entering wildlife food chains. J Wildlife Manage 71(1):103–108

Plessas IN, Volk HA, Kenny PJ (2013) Migraine-like episodic pain behavior in a dog: can dogs suffer from migraines? J Veterinary Intern Med 27(5):1034–1040

Pryor PA, Hart BL, Bain MJ, Cliff KD (2001) Causes of urine marking in cats and effects of environmental management on frequency of marking. Journal of the American Veterinary Medical Association 219(12):1709–1713

Sherman BL, Mills DS, (2008) Canine Anxieties and Phobias: An Update on Separation Anxiety and Noise Aversions. Veterinary Clinics of North America: Small Animal Practice 38 (5):1081-1106

Stepien RL, Rapoport GS, Henik RA et al (2003) Comparative diagnostic test characteristics of oscillometric and Doppler ultrasound methods in the detection of systolic hypertension in dogs. J Veterinary Intern Med 17(1):65–72

Tisdall PLC, Hunt GB, Youmans KR, Malik R (2000) Neurological dysfunction in dogs following attenuation of congenital extrahepatic portosystemic shunts. Journal of Small Animal Practice 41(12):539–546

Truvé K, Dickinson P, Xiong A, York D, Jayashankar K, Pielberg G et al (2016) Utilizing the dog genome in the search for novel candidate genes involved in glioma development—genome wide association mapping followed by targeted massive parallel sequencing identifies a strongly associated locus. PLoS Genet 12(5)

Wolschrijn CF, Mahapokai W, Rothuizen J, Meyer HP, van Sluijs FJ (2011) Gauged attenuation of congenital portosystemic shunts: results in 160 dogs and 15 cats. Veterinary Quart pp 94–98

Yoshino T, Uchida K, Tateyama S, Yamaguchi R, Nakayama H, Goto N (1996) A retrospective study of canine senile plaques and cerebral amyloid angiopathy. Veterinary Pathol Online 33(2):230–234

Zeira O, Asiag N, Aralla M, Ghezzi E, Pettinari L, Martinelli L, Zahirpour D, Dumas PM, Lupi D, Scaccia S, Konar M, Cantile C (2015) Adult autologous mesenchymal stem cells for the treatment of suspected non-infectious inflammatory diseases of the canine central nervous system: safety, feasibility and preliminary clinical findings. J Neuroinflamm 12(181)

Phenotypic Variability and Clinical Staging of Canine Dementia

3

Aladár Mad'ari, Petr Novak, and Norbert Žilka

Cognitive dysfunction syndrome (CDS) or canine dementia represents a serious health problem for aged dogs, regardless of breed. CDS is characterised by deficits in learning, memory and spatial awareness, as well as changes to social interaction and sleeping patterns. Several studies, using owner-based observational questionnaires, have been performed to assess the severity of the disease or to identify first clinical symptoms before the onset of full-fledged dementia. Questionnaires include a broad range of items measuring appetite, drinking behaviour, barking, elimination behaviour, day/night rhythms, aimless behaviour, adaptive capabilities, social behaviour, perceptual ability, disorientation, memory, and personality changes. It is important to note that canine dementia probably does not represent a single disease entity; rather, it may have various phenotypic presentations. Generally, the cognitive impairment and other clinical features of CDS gradually worsen as the disease progresses. It is generally accepted that at least three main stages of the disease—mild, moderate, and severe—can be recognised. Despite this fact, there is no consensus regarding the thresholds for discrimination of various stages. This chapter will guide the reader through the current knowledge on the clinical variability and staging of canine dementia.

A. Mad'ari (✉)
University of Veterinary Medicine and Pharmacy in Kosice,
Komenskeho 73, Kosice, Slovakia
e-mail: aladar_madari@yahoo.com

P. Novak
Institute of Neuroimmunology, Slovak Academy of Sciences,
Dubravska 9, Bratislava, Slovakia
e-mail: petr.novak@savba.sk

N. Žilka
Institute of Neuroimmunology, Slovak Academy of Sciences,
Dubravska 9, Bratislava, Slovakia

Institute of Neuroimmunology n.o., Dvorakovo nabrezie 45, Bratislava, Slovakia
e-mail: norbert.zilka@savba.sk

G. Landsberg et al. (eds.), *Canine and Feline Dementia*,
DOI 10.1007/978-3-319-53219-6_3

3.1 The Clinical Picture of Canine Dementia in a Nutshell

The cognitive decline and behavioural presentation of canine cognitive dysfunction syndrome (CDS) indicates similarity in dementia syndromes between pets and humans (Landsberg and Araujo 2005). The classic clinical signs of CDS are disorientation, interaction changes (such as decline in response to learned commands), sleep and wakefulness cycle changes, change in hygiene habits and house soiling, and activity changes (DISHA) (Landsberg et al. 2012). A drop in capability and adaptability is evident, generally accompanied by increased rates of anxiety and repetitive behaviour (Gunn-Moore 2011). Unambiguous identification of behavioural characteristics for cognitive decline can allow early intervention, delaying progression of the disease (Dowling and Head 2012).

3.1.1 Disorientation and Activity Changes

Demented dogs display aimless wandering and pacing, which, along with a reduction of spatial orientation and growing confusion, can cause the animal to become trapped in corners or behind furniture (Gonzalez-Martinez et al. 2011). A commonly seen problem is distinguishing between internal and external doors, such as going to the wrong side of the door upon arrival and departure (Landsberg and Araujo 2005). Dogs often ask to go out, but, once outside, disorientation becomes apparent once the impaired animal is a short distance from its home, manifesting in anxiety and confusion about the return route (Landsberg et al. 2012).

3.1.2 Changes in Social Interactions

Social interaction between dog and man is one of the pivotal aspects where canine dementia becomes apparent (Madari et al. 2015). Interactions of the dog with family members become less frequent and/or less intense. Social behaviour abnormalities become apparent in the expression of greetings or in association with feeding behaviours. Greeting behaviour patterns are incomplete in their expression, with pets showing disorientation in response to owners' attempts to interact with them or repeating greeting behaviour without previous separation time. Dogs may appear confused or even frightened by welcoming rituals. Dogs may beg for food but refuse to eat, begging to be fed once more a few minutes later. Behavioural patterns lasting a few minutes, sometimes hours or even the whole day, are repeated, which is especially noticeable with vocalisation and destructive behaviours. These social changes are manifestations of confusion and anxiety, similar to human dementia (Gunn-Moore et al. 2007). Confusion is considered to be one of the main features of CDS (Azkona et al. 2009).

3.1.3 Sleep and Wakefulness Cycle Changes

Sleep-wake cycle changes may be observed especially in later stages of the disease. Dogs might sleep more during the day but are active during the night. These changes are often associated with excessive anxiety-driven vocalisation, such as barking or howling. Demented dogs may wander, pace, or even scratch the ground. This behaviour is especially burdensome to owners and their surroundings (Landsberg and Araujo 2005; Landsberg et al. 2012).

3.1.4 Changes in Hygiene Habits

The loss of learned behaviours, such as toilet training, represents the most common sign attributable to CDS. Uncontrolled indoor elimination occurs in the absence of apparent medical problems or environmental changes (such as lack of access to an appropriate area) that would prevent the dog from following its toilet training. The dog may eliminate even when the owner is present. Soiling may occur in random locations (Neilson et al. 2001; Azkona et al. 2009; Landsberg and Araujo 2005).

To summarise, CDS represents a disorder with a very broad spectrum of clinical symptoms, ranging from memory impairment to changes in hygiene habits. These changes profoundly disturb the bond between dog and human by diminishing meaningful interaction between them (whether it was for the purpose of companionship or for professional use) on one hand and, on the other hand, by increasing the incidence of unmanageable disturbing behaviour (such as house soiling) (Osella et al. 2007; Salvin et al. 2011; Landsberg et al. 2012).

3.2 Questionnaires: A Double-Edged Sword in the Diagnostics of CDS

A variety of scales and questionnaires are used to diagnose and evaluate cognitive decline in pets. Generally, they constitute useful tools for the acquisition of information about the cognitive function of aged dogs and prevalence of cognitive decline in the domestic canine population. Many different questionnaires have been created (Colle et al. 2000; Neilson et al. 2001; Osella et al. 2007; Azkona et al. 2009, Salvin et al. 2011; Landsberg et al. 2012; Madari et al. 2015) since initial efforts by Ruehl et al. (1995). Questionnaires include a broad range of items measuring appetite, drinking behaviour, barking, elimination behaviour, day/night rhythms, aimless behaviour, adaptive capabilities, social behaviour, perceptual ability, disorientation, memory and personality changes (Table 3.1).

One of the main weaknesses of the questionnaires is the subjective evaluation of the dog's cognitive status made by a pet owner (Salvin et al. 2011). The use of the pet owner as untrained evaluator of behavioural changes represents an impactful source of inaccurateness. Several extensive studies on CDS were based either on phone

Table 3.1 The list of questionnaires for canine cognitive dysfunction

Questionnaire/items	Colle et al. (2000)	Neilson et al. (2001)	Osella et al. (2007)	Azkona et al. (2009)	Golini et al. (2009)	Salvin et al. (2011)	Landsberg et al. (2012)	Rosado et al. (2012)	Fast et al. (2013)	Madari et al. (2015)
Activity	–	–	+	–	+	+	+	–	+	-
Aggression	–	–	+	–	–	–	+	–	+	+
Anxiety	–	–	–	–	+	–	+	–	+	-
Commands	+	–	+	+	+	–	+	+	+	+
Drinking	+	–	–	–	–	–	–	–	+	-
Eating	+	–	+	–	–	+	+	–	+	-
Elimination	+	+	+	–	+	–	+	+	–	+
Orientation	–	+	+	+	+	+	+	+	+	+
Sleep-wake cycle	+	+	+	+	+	–	+	+	+	+
Social interaction	+	+	+	+	+	+	+	+	+	+
Repetitive behaviour	+	–	+	–	+	+	+	+	+	+

+ Item from this category is in the questionnaire. – No item from this category is in the questionnaire consultations with dog owners (Neilson et al. 2001) or on distribution of a questionnaire via online or hard copy formats (Salvin et al. 2011). While the advantage of these approaches lies in large datasets, the main limitations of this approach are a lack of systematic control for potential impact of other disorders on the cognitive decline, and the subjective nature of evaluations performed by pet owners (Madari et al. 2015).

Another issue that should be taken into consideration is how individual items addressed by the questionnaire are quantified. Some authors prefer to use item scoring that reflects the degree of abnormal behaviour (Colle et al. 2000; Pugliese et al. 2005), while others aim to quantify the frequency of their occurrence using either four- or five-point frequency scales (Osella et al. 2007; Salvin et al. 2011). Salvin et al. (2011) proposed a 5-point scale based on the frequency of abnormal behaviour: 1 point—never, 2 points—once a month, 3 points—once a week, 4 points—once a day and 5 points—more than once a day. Similarly, we have proposed a 5-point scale for easy evaluation of behaviour: 0 point—abnormal behaviour of the dog was never observed, 2 points—abnormal behaviour of the dog was detected at least once within the last 6 months, 3 points—abnormal behaviour appeared at least once per month, 4 points—abnormal behaviour was seen several times per month, and 5 points—abnormal behaviour was observed several times a week (Madari et al. 2015).

To sum up, questionnaires serve as valuable tool for detection of disease severity and are a helpful screening tool for identification of early behavioural changes. Finally, they can quantify the level of impairment of various cognitive domains.

3.3 CDS Clinical Staging

As mentioned above, CDS may progress through several stages. A few studies have focused on clinical staging of cognitive decline. Based on the level of cognitive impairment, Pugliese et al. (2005) divided dogs into three groups: normal, light cognitive deficits and severe cognitive deficits. Unfortunately, this approach does not reflect the frequency of abnormal behaviour. Azkona et al. (2009) classified cognitive impairment into mild (one behavioural domain), moderate (two domains) or severe (three or four domains). This classification could be sometimes misleading, because the number of impaired categories does not necessarily precisely reflect the degree of cognitive decline. We found that two and sometimes three categories of cognition can be slightly impaired already in cognitively normal or mildly affected dogs (see Table 3.2).

We have proposed criteria for discrimination of three stages of the disease: mild cognitive impairment, moderate cognitive impairment, and severe cognitive impairment (the last stage being full-fledged canine dementia). For evaluation of the severity of the disease, we have used a universally applicable scale for diagnostics of the canine cognitive dysfunction syndrome—CADES—which contains 17 items distributed into four domains (spatial orientation, social interactions, sleep-wake cycles, and house soiling) related to changes in dogs' behaviour. Staging criteria for

Table 3.2 Stages of cognitive decline in dogs, as defined by questionnaires

Questionnaires	Stages
Neilson et al. (2001)	Mild impairment
	Severe impairment
Azkona et al. (2009)	Mild impairment
	Moderate impairment
	Severe impairment
Golini et al. (2009)	Dogs with CDS
	Dogs without CDS
Salvin et al. (2011)	Normal ageing
	Query CCD
	Dementia
Landsberg et al. (2012)	Unimpaired
	Impaired
	Severely impaired
Rosado et al. (2012)	Mild cognitive impairment
	Moderate cognitive impairment
	Severe cognitive impairment
Fast et al. (2013)	Normal cognitive status
	Borderline CCD
	CCD
Madari et al. (2015)	Normal ageing
	Mild cognitive impairment
	Moderate cognitive impairment
	Severe cognitive impairment

CCD canine cognitive dysfunction, *CDS* cognitive dysfunction syndrome

detection of the severity of disease were based on the severity of behavioural changes, the number of affected domains, and total scores (Madari et al. 2015).

We have described *normal ageing* as a clinical status when dog owners usually do not notice any apparent changes in their pets' behaviour. The number of affected domains varies from 0 to 2. Early behavioural changes appear occasionally (Table 3.3).

Similarly, in the case of *mild cognitive impairment*, pet owners often do not recognise any changes in their pets' behaviour. Evaluation and interview by a veterinary clinician can identify very mild changes of the dog's behaviour, particularly alterations in interaction with owners or other pets, reduction in activity during the day and increased activity at night and, rarely, inappropriate indoor elimination. The number of slightly affected domains varies from 2 to 4.

When CDS progresses into the stage of *moderate cognitive impairment*, pet owners may observe salient behavioural changes, such as undesirable indoor elimination (urination, defaecation) or hyperactivity through the night. At this stage, dogs require more care than ever before. The number of affected domains varies from 2 to 4.

Table 3.3 List of questionnaires for canine cognitive dysfunction and their characteristics

Questionnaire/ characteristic	Colle et al. (2000)	Osella et al. (2007)	Azkona et al. (2009)	Golini et al. (2009)	Salvin et al. (2011)	Landsberg et al. (2012)	Rosado et al. (2012)	Fast et al. (2013)	Madari et al. (2015)
1. Fill-out time	15 min	20 min	10 min	15 min	5 min	15 min	15 min	15 min	10 min
2. Filled by an owner	–	–	–	–	+	±[a]	–	–	–
3. Filled by a specialist	+	+	+	+	–	±[a]	+	+	+
3a. A dog was presented during filling out a questionnaire	+	+	–	+	–	±[a]	+	±	+
4. Special examination	–[b]	+	–[c]	+	–	±[a]	+	+	+
4a. Clinical examination	–[b]	+	–[c]	+	–	±[a]	+	+	+
4b. Behavioural examination	–[b]	–	–[c]	+	–	±[a]	+	–	+
4c. Differential diagnostic	–[b]	+	–[c]	–	–	±[a]	+	–	+
5. Number of items	10	39	15	32	13	33	22	30	17
6. Dog age	3–19y	>7y	≥9y	≥7y	>8y	–	≥9y	>8y	≥8y

min minutes, *y* years

[a]Landsberg et al. (2012) did not state questionnaire characteristics, but they recommended performing a special examination of the dog before every testing

[b]Dogs were not specially examined, because Colle et al. (2000) examined dogs before euthanasia was requested by the owner

[c]Azkona et al. (2009) performed phone interviews with dog owners, but dogs were chosen based on medical history in patient database

In the final stage, *severe cognitive impairment* or canine dementia, pet owners report severe detrimental changes in behaviour, which markedly impair the quality of the coexistence between owner and dog. All four domains were affected, some of them heavily. A majority of animals shows apparent decline (over 10 points per domain) either in three or four domains. The CADES score was higher than 45 points.

3.4 The Phenotypic Variability of CDS

CDS encompasses a broad spectrum of clinical symptoms, and not all of them are present in all senior dogs. The variability in the clinical presentation is most probably caused by selective damage to various brain areas. Variations in phenotype were described in some studies (Madari et al. 2015; Schutt et al. 2015). Although the authors used different questionnaires, both studies demonstrated the existence of a range of individual phenotypic patterns. In our study, we found that 57.1% of dogs suffering from moderate cognitive impairment had severely impaired social interactions and sleep-wake cycles, 26% were impaired only in social interactions, 8.2% displayed impairment of both spatial orientation and social interaction, and 8.2% showed impairment exclusively in sleep-wake cycles. Of the 38 dogs suffering from severe cognitive dysfunction included in the study, 26.7% demonstrated concurrent impairment in spatial orientation, social interaction and sleep-wake cycles, and 26.7% in all four domains. Our results showed that social interactions and sleep-wake cycles were the most impaired categories (Madari et al. 2015). These findings indicate CDS is a multifarious disorder affecting various behavioural domains; with increasing severity of the disorder, more domains become affected.

The variety of clinical manifestations can be explained by variation in the distribution of pathological changes in different brain areas and selective vulnerability of said areas. From a plethora of studies on human dementia patients (Braak and Braak 1991; Neary et al. 1998; Mackenzie et al. 2010; Josephs et al. 2011), we know that the frontal and temporal cortices, as well as the hippocampus and entorhinal cortex represent the brain areas that are most vulnerable to neurodegeneration. Further studies are warranted to explore the direct relationship between the extent of the damage in specific brain areas and phenotypic patterns of canine dementia.

3.5 Summary

Canine dementia constitutes an unmet medical need. The number of dogs suffering from dementia rises quickly because we have prolonged the lifespan of dogs and cats. Current diagnostic approaches utilise various forms of questionnaires that can identify the disease in early stages of development. CDS passes through several stages from mild, moderate to severe cognitive impairment. CDS seems to have a slightly variable phenotype, which may differ between demented dogs. Taking these findings into consideration allows therapy to be started in the earliest stages of the disease.

Acknowledgement The work was funded by APVV-15-0613 and APVV-14-0872 and VEGA grant 2/0164/16.

References

Azkona G, Garcia-Belenguer S, Chacon G, Rosado B, Leon M, Palacio J (2009) Prevalence and risk factors of behavioural changes associated with age-related cognitive impairment in geriatric dogs. J Small Animal Pract 50(2):87–91

Braak H, Braak E (1991) Neuropathological staging of Alzheimer-related changes. Acta Neuropathol 82(4):239–259

Colle MA, Hauw JJ, Crespeau F, Uchihara T, Akiyama H, Checler F, Pageat P, Duykaerts C (2000) Vascular and parenchymal A beta deposition in the aging dog: correlation with behavior. Neurobiol Aging 21:695–704

Dowling ALS, Head E (2012) Antioxidants in the canine model of human aging. Biochimica Et Biophysica Acta-Mol Basis Dis 1822(5):685–689

Fast R, Schutt T, Toft N, Moller A, Berendt M (2013) An observational study with long-term follow-up of canine cognitive dysfunction: clinical characteristics, survival, and risk factors. J Veterinary Intern Med 27(4):822–829

Golini L, Colangeli R, Tranquillo V, Mariscoli M (2009) Association between neurologic and cognitive dysfunction signs in a sample of aging dogs. J Veterinary Behav-Clin Appl Res 4:25–30

Gonzalez-Martinez A, Rosado B, Pesini P, Suarez ML, Santamarina G, Garcia-Belenguer S, Villegas A, Monleon I, Sarasa M (2011) Plasma beta-amyloid peptides in canine aging and cognitive dysfunction as a model of Alzheimer's disease. Exp Gerontol 46(7):590–596

Gunn-Moore DA (2011) Cognitive dysfunction in cats: clinical assessment and management. Topics Companion Animal Med 26(1):17–24

Gunn-Moore D, Moffat K, Christie LA, Head E (2007) Cognitive dysfunction and the neurobiology of ageing in cats. J Small Animal Pract 48(10):546–553

Josephs KA, Hodges JR, Snowden JS, Mackenzie IR, Neumann M, Mann DM, Dickson DW (2011) Neuropathological background of phenotypical variability in frontotemporal dementia. Acta Neuropathol 122:137–153

Landsberg G, Araujo JA (2005) Behavior problems in geriatric pets. Veterinary Clin N Am-Small Animal Pract 35(3):675–698

Landsberg GM, Nichol J, Araujo JA (2012) Cognitive dysfunction syndrome a disease of canine and feline brain aging. Veterinary Clin N Am-Small Animal Pract 42(4):749–768

Mackenzie IR, Neumann M, Bigio EH, Cairns NJ, Alafuzoff I, Kril J, Kovacs GG, Ghetti B, Halliday G, Holm IE, Ince PG, Kamphorst W, Revesz T, Rozemuller AJ, Kumar-Singh S, Akiyama H, Baborie A, Spina S, Dickson DW, Trojanowski JQ, Mann DM (2010) Nomenclature and nosology for neuropathologic subtypes of frontotemporal lobar degeneration: an update. Acta Neuropathol 119:1–4

Madari A, Farbakova J, Katina S, Smolek T, Novak P, Weissova T, Novak M, Zilka N (2015) Assessment of severity and progression of canine cognitive dysfunction syndrome using the CAnine DEmentia Scale (CADES). Appl Animal Behav Sci 171:138–145

Neary D, Snowden JS, Gustafson L, Passant U, Stuss D, Black S, Freedman M, Kertesz A, Robert PH, Albert M, Boone K, Miller BL, Cummings J, Benson DF (1998) Frontotemporal lobar degeneration: a consensus on clinical diagnostic criteria. Neurology 51:1546–1554

Neilson JC, Hart BL, Cliff KD, Ruehl WW (2001) Prevalence of behavioral changes associated with age-related cognitive impairment in dogs. J Am Veterinary Med Assoc 218(11):1787–1791

Osella MC, Re G, Odore R, Girardi C, Badino P, Barbero R, Bergamasco L (2007) Canine cognitive dysfunction syndrome: prevalence, clinical signs and treatment with a neuroprotective nutraceutical. Appl Animal Behav Sci 105(4):297–310

Pugliese M, Carrasco JL, Andrade C, Mas E, Mascort J, Mahy N (2005) Severe cognitive impairment correlates with higher cerebrospinal fluid levels of lactate and pyruvate in a canine model of senile dementia. Progr Neuro-Psychopharmacol Biol Psychiatry 29:603–610

Rosado B, Gonzalez-Martinez A, Pesini P, Garcia-Belenguer S, Palacio J, Villegas A, Suarez ML, Santamarina G, Sarasa M (2012) Effect of age and severity of cognitive dysfunction on spontaneous activity in pet dogs—part 1: locomotor and exploratory behaviour. Veterinary J 194(2):189–195

Ruehl WW, Bruyette DS, DePaoli A, Cotman CW, Head E, Milgram NW, Cummings BJ (1995) Canine cognitive dysfunction as a model for human age-related cognitive decline, dementia and Alzheimer's disease: clinical presentation, cognitive testing, pathology and response to 1-deprenyl therapy. Curr Neurochem Pharmacol Aspects of Biogenic Amines 106:217–225

Salvin HE, McGreevy PD, Sachdev PS, Valenzuela MJ (2011) The canine cognitive dysfunction rating scale (CCDR): a data-driven and ecologically relevant assessment tool. Veterinary J 188:331–336

Schutt T, Toft N, Berendt M (2015) Cognitive function, progression of age-related behavioral changes, biomarkers, and survival in dogs more than 8 years old. J Veterinary Intern Med 29(6):1569–1577

The Aged Dog as a Natural Model of Alzheimer's Disease Progression

4

Joseph A. Araujo, Jessica Baulk, and Christina de Rivera

Alzheimer's disease is the most common cause of dementia in humans and is rapidly becoming an unmet medical need of epidemic proportions. More than 50 million people are expected to be diagnosed with Alzheimer's disease by 2050 unless efficacious disease-modifying treatments are developed. Although billions of dollars have been allocated to this cause, more than 190 putative Alzheimer's disease drugs have failed in the clinic, exemplifying the high risk to reward ratio of Alzheimer's disease therapeutic development. One factor implicated in this high failure rate is the limitation of animal models to accurately predict clinical outcomes.

Animal models are essential for demonstrating potential target relevance and target engagement of novel therapeutics. The fact that some of these models, such as transgenic mice, have a 100% failure rate for predicting clinical outcomes of putative drugs exemplifies their limitations and the need for additional models that better recapitulate the multifactorial nature of Alzheimer's disease progression.

Aged dogs naturally develop Alzheimer's-like neuropathological changes, as well as cognitive domain-specific impairments consistent with early stages of Alzheimer's disease progression. Moreover, cross-sectional data in differentially aged dogs suggests that Alzheimer's relevant biomarker changes are also found in dogs and could be used to monitor early-stage Alzheimer's-like disease progression. The naturalistic progression of these changes in canine aging has not been linked to a single target, which suggests aged dogs may be used as a confirmatory animal model for preclinically evaluating putative Alzheimer's therapeutics. In this respect, the aged dog model accurately predicted the clinical outcome of both the gold

J.A. Araujo (✉)
InterVivo Solutions, Toronto, ON, Canada
e-mail: josepha@intervivo.com

J. Baulk • C. de Rivera
Vivocore, Inc., Toronto, ON, Canada
e-mail: jessicab@vivocore.com; christinad@vivocore.com

G. Landsberg et al. (eds.), *Canine and Feline Dementia*,
DOI 10.1007/978-3-319-53219-6_4

standard symptomatic Alzheimer's disease therapeutic, donepezil, and the failure of active fibrillary amyloid vaccination strategies.

The aged dog, therefore, provides a valuable preclinical animal model for assessing Alzheimer's disease therapeutics with demonstrable translational value for predicting clinical outcomes of both symptomatic and disease-modifying Alzheimer's disease interventions. Given the translational validity of the model, it is likely that Alzheimer's disease clinical research could also be used for improving care of senior dogs.

4.1 Introduction

Alzheimer's disease (AD) is a progressive and chronic neurodegenerative disease and the major cause of dementia in the elderly, accounting for 50–75% of all dementia cases (Alzheimer 1906; Grossman et al. 2006; Qiu et al. 2009; Selkoe 2001; Uzun et al. 2011). According to diagnostic criteria guidelines (McKhann et al. 1984), AD onset occurs between 40 and 90 years of age, with onset being highest in populations greater than 65 years of age; early-onset cases, associated with dominant familial gene mutations, represent less than 5% of all AD cases (Ott et al. 1995; Selkoe 2001). Current global estimates of AD prevalence range from 15 to 27.7 million cases in the general population, but are anticipated to rise to greater than 115 million cases by 2015, indicating there is unmet medical need of epidemic proportions (WHO 2012).

Longitudinal clinical studies aimed at characterizing the progression of AD in aged human populations suggest that neuropathological changes develop decades prior to clinical diagnosis, which can only be confirmed by postmortem confirmation of both pathological hallmarks of AD—senile plaques and neurofibrillary tangles. Therefore, it is increasingly accepted that there is a substantial prodromal phase of AD, and that disease progression, as well as therapeutic efficacy, could be monitored using pathophysiological biomarkers, which provide a paradigm for targeting patient populations during earlier, and presumably more responsive, stages of the disease.

Although an enormous amount of financial resource has been invested in developing AD therapeutics, more than 190 have failed in human clinical trials, and only four drugs have been approved (Becker and Greig 2012; Cummings et al. 2014). Therefore, there is a significant unmet need for novel, and efficacious, AD-modifying therapeutics that either halt or slow progression to clinical AD (Klafki et al. 2006). The failure to develop efficacious drugs for AD can be attributed in part to translational limitations of preclinical animal models to predict clinical outcomes. The current chapter aims to demonstrate that the aged dog model naturally models several characteristics of AD and can be used to evaluate the potential efficacy of putative therapeutics across varying stages of AD progression. Specifically, the aged dog is a translationally relevant preclinical animal model because the model accurately predicts outcomes of human clinical trials.

4.2 Alzheimer's Disease

Alzheimer's disease is a neurodegenerative disease originally described by Alois Alzheimer (Alzheimer 1906) in which patients present with progressive cognitive deficits across multiple cognitive domains that ultimately result in dementia and death. Although the cause of the disease is unknown, there are both distinguishing and associated neuropathological changes present in the disease. More recently, biomarkers of these pathophysiological changes are being investigated in the effort to diagnose disease in early stages of progression, and potentially to monitor disease-modifying interventions. The current section focuses on these three key features–neuropathology, cognitive deficits, and associated biomarkers of pathophysiological progression.

4.2.1 Neuropathological Features

The hallmark neuropathological features of AD are extracellular senile, or neuritic, plaques and intracellular neurofibrillary tangles (NFT) in the neocortex and temporal lobe structures (Hyman and Trojanowski 1997). The senile plaque pathology is a result of extracellular focal deposition of amyloid beta (Aβ) protein, and the NFT pathology is comprised of intracellular paired helical filaments and straight filaments of abnormally, or hyper-, phosphorylated tau protein (Markesbery 2010). These two pathologies can occur decades prior to clinical manifestation of the disease, and are accompanied by extensive neurodegenerative processes that include neuronal and synaptic dysfunction and loss, leading to widespread cortical atrophy and alterations in multiple neurotransmitter systems and neural networks (Jack et al. 2010).

In addition to these hallmark pathological features, several other neuropathological changes are associated with AD, including decline in neurotransmitter systems, particularly the cholinergic system (Bartus 2000; Bartus et al. 1982; Norbury et al. 2005; Picciotto and Zoli 2002; Reinikainen et al. 1987, 1990; Rossor and Iversen 1986), extensive cortical and/or brain region-specific atrophy measured by both postmortem evaluation and in vivo magnetic resonance imaging (MRI) (Double et al. 1996; Fox and Freeborough 1997; Jack et al. 2005), and increased oxidative stress (Markesbery 1997; Smith et al. 2000). Thus, the two neuropathological diagnostic criteria for an AD diagnosis represent only a part of the neuropathological processes occurring in AD progression.

4.2.2 Cognitive Decline

The clinical diagnosis of AD is based on a progressive dementia with cognitive impairment in multiple independent cognitive domains and behavioral (neuropsychiatric) changes (Albert 2011; Marin et al. 1997; McKhann et al. 1984, 2011) that ultimately result in death (Cummings 2000). In the majority of cases, deficits in episodic memory occur in conjunction with changes in at least one other cognitive domain, although other variations of cognitive deficits are described (Albert 2011; McKhann et al. 2011). Moreover, episodic memory deficits are reported in subpopulations with mild cognitive impairment (MCI) who are at risk for development of AD; however, this subgroup does not differ cognitively from normal aged-matched controls on other cognitive measures, which suggests that deficits in episodic memory may precede clinical diagnosis of AD (Sperling et al. 2011).

The most devastating feature of AD is perhaps the progression of cognitive decline leading to overt dementia, in which multiple cognitive domains are affected over several years until individuals are no longer able to live independently (Albert 2011). Cognitive domains most commonly affected in the pathophysiological progression of AD include memory, executive function/attention, visuospatial skills, and language (Albert et al. 2011). The majority of AD patients show a relatively consistent pattern of cognitive decline, which is referred to as progressive amnestic disorder, and is generally characterized by early impairments in episodic memory (Albert 2011; Backman et al. 2000; Grober et al. 1999; Linn et al. 1995; Perry et al. 2000; Petersen 1998; Petersen et al. 1994) consistent with early neuropathological changes in the entorhinal cortex and hippocampal formation (Braak and Braak 1991; Jack et al. 1992, 1997). Executive function deficits precede language and visuospatial deficits in a majority of AD cases (Amieva et al. 2004; Foldi et al. 2002; Grady et al. 1988; Lafleche and Albert 1995; Perry et al. 2000; Reid et al. 1996), although some report that visuospatial deficits also occur early in the course of AD progression (Becker et al. 1988; Johnson et al. 2009). Based on performance across various cognitive tasks, it is suggested that AD progression can be divided into amnestic cases, in which declarative memory deficits are most evident; non-memory or dysexecutive cases, in which deficits on executive function tasks are most evident with sparing of declarative memory; or multi-domain, or mixed, cases in which deficits across multiple cognitive domains are evident, including combined memory and language deficits (Delano-Wood et al. 2009; Eppig et al. 2012; Libon et al. 2010, 2011; Petersen 2004; Petersen and Morris 2005; Ritchie and Touchon 2000).

4.2.3 Pathophysiological Biomarkers of Progression

It is increasingly accepted that pathophysiological biomarkers can provide diagnostic and prognostic information for improved clinical management of patients in early stages of AD progression (Buerger et al. 2006; Hampel et al. 2008; Jack et al. 2010; Shaw et al. 2007). To a large extent, this has been facilitated by large longitudinal clinical trials such as the Alzheimer's Disease Neuroimaging Initiative (ADNI) and the Dominantly Inherited Alzheimer's Network (DIAN), which collect longitudinal data in patients with AD, mild cognitive impairment, and elderly controls with the aim of understanding disease progression (Jack et al. 2008; Morris et al. 2012; Moulder et al. 2013; Weiner et al. 2013). The emerging data indicate that AD processes begin decades before clinical manifestation, and to this end, Jack et al. (2010) propose a trajectory of neuropathological events progressing through cerebral amyloidosis, neuronal injury and synaptic dysfunction, and neurodegeneration, with memory impairment occurring prior to clinically diagnosed dementia. Specifically, markers of amyloidosis including increased amyloid positron emission tomography (PET) tracer retention and reduced cerebrospinal fluid (CSF) $A\beta_{42}$ are detected early in the disease process in patients that are clinically normal (de Leon et al. 2001; Jagust et al. 2006; Petrie et al. 2009). Neuronal and synaptic dysfunction can be monitored by increased CSF tau and decreased fluorodeoxyglucose uptake on PET imaging, respectively. Neurodegeneration can be identified by magnetic resonance imaging (MRI) measures of cerebral atrophy (Bobinski et al. 1999; Frisoni et al. 2010; Ramani et al. 2006; Scheltens et al. 2002). Moreover, combinations of biomarkers demonstrate improved ability to predict conversion to AD better than one marker alone (Vemuri et al. 2009a, b, 2010). Ultimately, biomarkers may serve as a method to track the in vivo efficacy of AD-modifying treatments in patient populations that are presymptomatic (Jack et al. 2010).

In addition to the abovementioned biomarkers, there are several other in vivo imaging, CSF, and blood-based biomarkers currently being explored. For example, metabolic markers of neuronal health can be measured by proton magnetic resonance spectroscopy (H-MRS). In patients with MCI or AD, levels of n-acetylaspartate (NAA) and myoinositol (INS) are of particular interest. NAA is an amino acid made in the mitochondria, which are located in neuronal cell bodies, axons, and dendrites (Birken and Oldendorf 1989; Simmons et al. 1991). Thus, NAA is thought to be a biomarker for neuronal loss. H-MRS studies have shown decreased NAA in MCI and AD patients (Kantarci et al. 2000; Rose et al. 1999). By contrast, INS is a glial metabolite. Kantarci et al. (2000) reported that INS levels are increased in MCI and AD patients. Given the large-scale longitudinal trials and the intensity of interest in novel biomarker of AD progression, it is likely that key biomarkers will be identified and validated in the near future.

4.3 The Aged Dog Model of Alzheimer's Disease Progression

Canine aging is associated with behavioral and neuropathological changes that parallel the progressive nature of human AD. Recent data examining in vivo imaging and CSF pathophysiological biomarkers in canine aging also demonstrate parallels to the biomarker changes observed in AD progression. The current section compares these three features of canine aging to AD progression to demonstrate the parallels of this naturally occurring disease process in dogs and humans.

4.3.1 Neuropathological Changes

Like AD patients, aged dogs also demonstrate amyloid plaques, and numerous studies have examined the composition and accumulation pattern of senile plaques in the aged canine brain (Cummings et al. 1996a, c; Mirra et al. 1991). This research is driven largely by the popularity of the amyloid hypothesis of AD that proposes that Aβ is a causative factor in disease progression (Hardy 2006). Mirra et al. (1991) found that senile plaques are composed of Aβ protein, which is identical in sequence between dogs and humans (Johnstone et al. 1991; Selkoe et al. 1987). The Aβ peptide is produced by the sequential cleavage of amyloid precursor protein (APP) by beta-secretase and gamma-secretase in both humans and dogs (Borghys et al. 2012; Murphy et al. 2010; Selkoe 1996), resulting in similar isoforms in both species (Portelius et al. 2010). In both dog and human, the 42-amino acid isoform ($A\beta_{42}$) is the predominant isoform found in the insoluble plaque deposits (Selkoe 2001; Wisniewski et al. 1990). The regional and temporal pattern of Aβ deposition in the human brain is well characterized (Braak and Braak 1991; Braak et al. 1993; Wisniewski et al. 1970). Similar to humans, Aβ deposition occurs earliest in the prefrontal cortex, and later in the temporal cortex, hippocampus, and occipital cortex of beagles (Giaccone et al. 1990; Head et al. 2000; Ishihara et al. 1991; Thal et al. 2002) and other breeds of dogs (Rofina et al. 2003, 2004, 2006; Smolek et al. 2016). In contrast to human plaque pathology, which includes both diffuse and dense core plaque, canine amyloid plaques are mainly of the diffuse subtype (Cummings et al. 1993; Giaccone et al. 1990; Morys et al. 1994; Okuda et al. 1994; Russell et al. 1992; Uchida et al. 1992). In the cerebrovasculature, the deposition of the 40-amino acid isoform of Aβ occurs in both dogs and humans, resulting in cerebrovascular amyloid angiopathy (Attems 2005; Attems et al. 2005; Herzig et al. 2006). Vascular Aβ deposits can result in disruption of the blood-brain barrier and vessel wall viability (Prior et al. 1996), as well as microhemorrhages (Uchida et al. 1991).

Several studies in beagle dogs indicate a positive correlation between cognitive impairments and increasing brain Aβ plaque load in addition to increasing age (Colle et al. 2000; Cummings et al. 1996a, c; Head et al. 1998; Rofina et al. 2006). In this respect, standardized neuropsychological tests (described in the subsequent section) have been useful for determining the impact of brain region-specific amyloid burden on cognitive domain-specific function independently of age. High levels of Aβ in the prefrontal cortex, for instance, are associated with reversal learning and complex working memory deficits, whereas high levels of Aβ in the entorhinal cortex are associated with impairments in size discrimination and reward- and object-approach learning (Cummings et al. 1996a; Head et al. 1998; Tapp et al. 2004); the prefrontal pathology-linked impairment is consistent with the known dependency of executive processes to prefrontal integrity. In pet dogs, the extent of Aβ plaque load also correlates significantly with behavioral changes independently of age (Colle et al. 2000; Rofina et al. 2006). On the other hand, impairments in complex working memory (measured by the delayed non-matching to position task (DNMP)) occur early in canine aging and likely precede amyloid plaque deposition (Studzinski et al. 2006). This pre-plaque impairment may be attributable to soluble amyloid oligomeric species, which are highly toxic, impair synaptic function, and correlate with cognitive dysfunction in humans (Lacor et al. 2004; Selkoe 2008; Tomic et al. 2009). Oligomeric amyloid also is increased with age in beagle dogs (Pop et al. 2012) and may serve as a target for AD-modifying therapeutics early in the disease process.

The second neuropathological hallmark of AD is NFT neuropathology, which generally is absent in aged animals, including dogs (Morys et al. 1994; Russell et al. 1992; Selkoe et al. 1987; Uchida et al. 1992); however, others report the presence of NFTs in aged dogs, albeit to a lesser extent than humans (Papaioannou et al. 2001; Schmidt et al. 2015). Regardless, several phosphorylated tau epitopes, as well as neuropil threads that are consistent with AD pathology in humans are found in the aged canine brain and may be linked to behavioral changes associated with cognitive dysfunction syndrome (Pugliese et al. 2006, Schmidt et al. 2015; Smolek et al. 2016; Yu et al. 2011). Therefore, it is unclear if tauopathy in dogs leads to synaptic and cell loss. Aged dogs often also show marked ventriculomegaly, thinning of the cerebral cortex, and reduced white matter (Gonzalez-Soriano et al. 2001; Kimotsuki et al. 2005; Su et al. 1998; Vandevelde et al. 2012). Moreover, brain atrophy varies by region; for example, prefrontal cortical volume decreases prior to hippocampal volume (Tapp et al. 2004). Significantly fewer neurons are found in the hilus of the dentate gyrus of aged dogs, and fewer Purkinje cells are found in the cerebellum of impaired dogs (Pugliese et al. 2007; Siwak-Tapp et al. 2008), which is consistent with the extensive neuronal loss in AD (Bobinski et al. 1997; West 1993). These reductions may reflect decreased neurogenesis in aged dogs (Pekcec et al. 2008; Siwak-Tapp et al. 2007). Another potential source of neurodegeneration is the increase in oxidative stress and reduced antioxidant capacity in canine aging, which is linked to cognitive deficits (Head et al. 2002; Hwang et al. 2008; Kiatipattanasakul et al. 1997; Opii et al. 2008; Papaioannou et al. 2001; Rofina et al. 2004, 2006; Skoumalova et al. 2003) and may be a result of age-related mitochondrial dysfunction (Head et al. 2009).

Although age-related modification in neurotransmitter systems has not been extensively investigated in the dog, there is growing evidence that alterations in neurotransmitter systems may be a cause of cognitive dysfunction and may be a result or cause of other AD-like neuropathological changes. Like humans, aged dogs are more sensitive to cognitive disruption caused by scopolamine than young dogs, which may be due in part to reduced muscarinic receptor numbers across multiple brain regions except the cerebellum (Araujo et al. 2011b; Reinikainen et al. 1987, 1990). Complex working memory performance is particularly susceptible to cholinergic disruption and is improved using cholinesterase inhibitors in both dogs and humans (Araujo et al. 2004, 2011a; Drachman and Leavitt 1974). Cognitively impaired dogs also show a significant reduction in the noradrenergic neurons in the locus coeruleus, which are also implicated in AD (Dringenberg 2000; Grudzien et al. 2007; Insua et al. 2010).

4.3.2 Cognitive Decline

Consistent with AD-like cognitive impairments, cognitive function tasks can be used to demonstrate that domain-specific cognitive decline which also occurs in canine aging (Araujo et al. 2008; Chan et al. 2002; Landsberg et al. 2012; Rofina et al. 2006; Siwak et al. 2001; Snigdha et al. 2012; Studzinski et al. 2006; Tapp et al. 2003). For example, memory and executive functions are impaired relatively early in the aging process (Studzinski et al. 2006; Tapp et al. 2001), and these impairments are positively correlated with brain region-specific amyloid burden and atrophy (Cummings et al. 1996c; Head et al. 1998; Rofina et al. 2006; Tapp et al. 2004). To objectively assess cognitive function across multiple cognitive domains in dogs, a number of validated laboratory-based neuropsychological tasks have been developed (Landsberg et al. 2012). These tasks are tester administered and utilize food rewards to reinforce correct responses.

Although the behavioral changes associated with canine cognitive dysfunction (CDS) are presumably due to brain neuropathology and impaired cognitive function (Landsberg et al. 2012, see Chaps. 1 and 5), early and subtle cognitive deficits may be difficult to identify in the home or clinical environment. Moreover, the behavioral changes associated with CDS generally occur later in life than the cognitive impairments detected using neuropsychological testing. In fact, impaired spatial working memory performance predicts behavioral changes that include altered sleep-wake cycles and decreased exploration and interaction (Siwak et al. 2001, 2003).

The test apparatus and neuropsychological tasks have been described extensively elsewhere (Landsberg et al. 2012), but here we describe a subset of tasks that are sensitive to both aging and therapeutic intervention. Importantly, these tasks can be used to examine therapeutic interventions both as a model for AD progression and as a preclinical model for CDS.

Delayed Non-matching to Position Task (DNMP): The DNMP task can be used to assess both complex visuospatial learning and short-term visuospatial working memory (Chan et al. 2002; Studzinski et al. 2006), although the former can only truly be tested once, limiting its use in assessing interventions. On each trial of the task, the dog is presented initially with a single object (e.g., white block) over one of the three possible locations and then is required to displace the object to obtain a food reward (Fig. 4.1a). The tray is then removed from the dog's sight, and a delay is initiated. Following the delay, the animal is then presented with two objects identical to that used in the original presentation. The dog is required to displace the object in the new location and receives a reward for responding correctly (see Fig. 4.1b).

The delay between the first and second presentations may be varied to modify the memory demands of the task. Old dogs can be separated into three groups based on DNMP performance—unimpaired, impaired, and severely impaired—which may be analogous to the various cognitive stages of AD progression (Adams et al. 2000). Also, learning and memory impairments on the DNMP can be detected as early as middle age (Studzinski et al. 2006). In terms of assessing learning, this is a particularly useful task because performance varies directly with age: young dogs learn the task rapidly, older dogs learn more slowly, and very old dogs may be unable to learn even with extensive training (Studzinski et al. 2006); specifically, age predicted 48.2% of the variability in learning the DNMP task. Beagle dogs ranging from 1 to 11.99 years generally made more errors with increasing age, and mild visuospatial deficits were detected by 6 years, which precedes the typical onset of Aβ accumulation in the dog brain by approximately 2 years. This suggests the DNMP task can serve as an early marker for cognitive decline in the dog, and that age-related

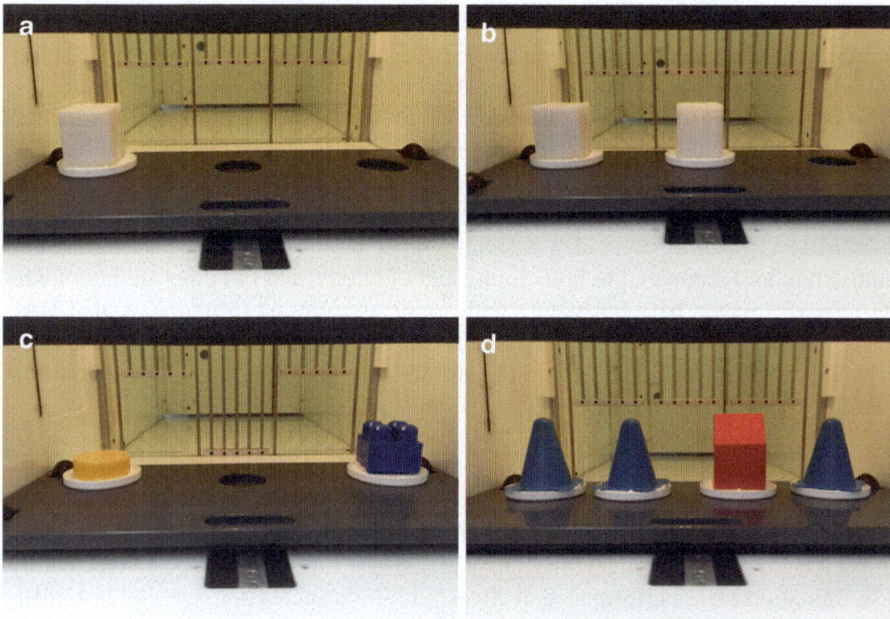

Fig. 4.1 (**a**) Sample phase of the DNMP task. The dog is required to displace a single block in one of the three positions to obtain a food reward. (**b**) Choice phase of the DNMP task. Following sample presentation and a predetermined delay, the dog must select the block in the new (non-match) position to obtain a food reward. (**c**) Simple discrimination learning and reversal task. During learning, the dog is required to continually select one of the two objects until learning criteria are reached. Subsequently, the reward contingency is reversed during reversal learning discrimination. (**d**) Selective attention task. The animal is required to select a single object and not a distractor; however, the number of distractors (i.e., negative stimuli) can vary from 0 to 3 during each trial

changes in visuospatial function in the dog models are seen in humans (Studzinski et al. 2006), possibly due to the influence of the cholinergic system on its performance (Araujo et al. 2004, 2011a, b).

Object Discrimination and Reversal Learning Tasks: Discrimination learning requires the animal to select one of the two different objects to obtain a food reward (Milgram et al. 1994). Objects may vary in color, shape, size, or a combination of the three (see Fig. 4.1c). Discrimination learning performance typically remains intact with age, possibly because an associative learning strategy can be used (Cotman et al. 2002). However, increasing difficulty of the task (by using more similar objects) exacerbates age-related deficits (Tapp et al. 2003). Once the task is learned, reversal learning can be tested by modifying the reward contingency such that the previously non-rewarded object is rewarded. Reversal learning is highly age sensitive in dogs, as the task predominately relies on executive function, requiring the subject to inhibit previously rewarded actions and shift responses to a new stimulus (Mongillo et al. 2013; Tapp et al. 2003). Specifically, both delayed learning and perseverative responding increase with increasing age.

Selective Attention Task: The selective attention task, or variable object discrimination task, is similar to discrimination learning in that the dog is required to select a single object for a reward; however, in this task, rather than one negative stimulus, there is a variable number of distractor objects (0 to 3) on any given trial (see Fig. 4.1d). With increasing distractor number, accuracy decreases significantly, and latency to respond increases across dogs of all ages (Snigdha et al. 2014), which is consistent with human selective attention impairments on tasks such as the visual search task (Parasuraman et al. 1995; Snigdha et al. 2012). Senior dogs are significantly impaired compared to both young and old dogs, and old dogs are intermediate in performance between young and senior age groups (Snigdha et al. 2014). These results suggest that aging impairs the ability of canines to discriminate between task-relevant and task-irrelevant stimuli, likely due to impairments in attentional processes related to inhibitory control and engagement.

The selective attention task employs a previously learned object to permit repeated testing in longitudinal aging or therapeutic intervention studies. A similar paradigm, oddity discrimination learning, examines learning when the incorrect object is present in duplicate and in which difficulty can be increased by using objects that are more similar (Milgram et al. 2002b; Tapp et al. 2003). On the oddity discrimination task, performance declines with both increasing age and task difficulty (Milgram et al. 2002b), presumably due to the increased attentional demands associated with increased distractor number (Araujo et al. 2008).

4.3.3 Pathophysiological Biomarkers

The longitudinal pattern of AD-relevant biomarker changes has not been well studied in the dog, but there is increasing evidence from cross-sectional studies suggesting biomarkers of brain amyloidosis, neuronal and synaptic dysfunction, and neurodegeneration, which parallel those seen in AD progression, are also evident in canine aging. Head et al. (2008) demonstrated that percent $A\beta_{42}$ CSF levels decline from middle to senior age in dogs, and that this decline is inversely correlated with increasing brain amyloid burden. This parallels the findings in humans that decreasing levels of $A\beta_{42}$ in CSF correlate with increased brain amyloid burden measured using amyloid-binding ligands with PET imaging (Fagan et al. 2006; Grimmer et al. 2009; Jagust et al. 2009; Sperling et al. 2011; Tolboom et al. 2009). However, the amyloid-binding ligands currently used for clinical or research purposes have not demonstrated specific amyloid binding of canine plaques, which may be due to the diffuse nature of the canine amyloid plaques (Cummings et al. 1996b; Czasch et al. 2006; Rofina et al. 2004; Torp et al. 2000a, b; Yoshino et al. 1996; Verhoeff et al. 2003).

More recently, we investigated changes in CSF levels of amyloid in 130 beagle dogs divided into four age groups (young, n = 17, 2.00–2.58 years; middle age, n = 21, 6.33–7.25 years; old, n = 57, 8.00–11.92 years; and senior, n = 35, 12.00–17.33 years) (Araujo et al. 2013a). A significant effect of age on percent $A\beta_{42}$ was found, which reflected significantly higher levels in middle-aged dogs compared to all other age groups. This finding supports the hypothesis that a threshold concentration of $A\beta_{42}$ is achieved during middle age prior to brain amyloid deposition (Sutphen et al. 2015). Percent $A\beta_{42}$ was significantly lower in senior dogs compared to old dogs, which is consistent with the findings of Head et al. (2008). We also examined CSF levels of total tau in a subset of these dogs and found a significant age-dependent effect on CSF total tau levels, which increased from middle age to old and from old to senior, analogously to findings in human AD (Buerger et al. 2006). The mean within-subject coefficient of variance was 9.9% for $A\beta_{42}$, 10.8% for $A\beta_{40}$, and 9.7% for total tau, suggesting CSF $A\beta_{42}$ and total tau levels may serve as reliable and translational biomarkers for preclinical aged canine studies investigating AD progression and possibly for predicting CDS.

Many in vivo imaging studies have been conducted in dogs, and several results parallel human AD findings. A pilot study examining fluorodeoxyglucose uptake on PET imaging in six adult and six senior dogs revealed significantly decreased fluorodeoxyglucose uptake across multiple brain regions consistent with the hypometabolism reported in AD patients (Araujo et al. 2013c). When cognitive tasks of varying difficulty were administered prior to imaging, brain region-specific engagement and disengagement were seen in adult, but not senior dogs, which suggests impairments in synaptic function and cortical processing (Horwitz et al. 1999). Magnetic resonance imaging studies demonstrate age-related ventriculomegaly as well as tissue volume loss of the frontal lobe, which occurs prior to atrophy of the hippocampus (Dimakopoulos and Mayer 2002; Su et al. 1998, 2005; Tapp et al. 2004). Moreover, a longitudinal MRI study in 47 dogs ranging from 8 to 11 years of age at the start of the 4-year study indicated that ventricular volume increased significantly over the last 2 years of the study and that the increase in ventricular volume was greater in older dogs than younger dogs; a 2.8% compared to 1.1% increase in ventricular volume was seen in dogs that were 11 compared to 8 years old at the start of the study, respectively (Su et al. 2005). Imaging studies using H-MRS also indicate an age-dependent reduction in NAA levels in dogs, which also is reported in AD patients (Adalsteinsson et al. 2000; de Rivera et al. 2007).

Therefore, biomarker patterns seen in AD progression are also found in canine aging, albeit the temporal sequence of these events has not been well studied in canine longitudinal studies. The same biomarkers may also prove valuable for predicting likely progression to CDS in pet dogs; however, the cost and invasiveness currently limit their use in the clinical setting. Regardless, biomarkers can be incorporated into laboratory canine studies designed to understand the progression of neuropathological events, their impact on cognitive function, and evaluation of novel therapeutics for AD-modifying drugs.

4.4 Predictive Validity of the Aged Dog Model

The validity of an animal model or test can be evaluated using various criteria which are beyond the scope of the current chapter. Rather, the ability of animal model to accurately predict clinical outcomes, or predictive validity, is the focus of the current section, which is pertinent given the absence of available AD-modifying therapeutic agents. In this respect, (pharmacological) predictive validity can be evaluated by determining how accurately an animal model predicts the results of human therapeutic clinical trials. While several animal models accurately predicted the symptomatic benefits of cholinesterase inhibitors on AD-like memory deficits, more than 190 compounds that were positive in animal models of AD have failed in clinical trials. This high number of false positives demonstrates a general lack of pharmacological validity in the most commonly used animal models, and transgenic mouse models, for example, have failed to accurately predict a positive result (Zahs and Ashe 2010). Although the absence of a gold standard AD-modifying therapeutics limits the ability to evaluate predictive validity for disease-modifying therapeutics, the dog has demonstrated ability to detect both true positives and false negatives (Studzinski et al. 2005) and also shows predictive value in non-pharmacological studies.

4.4.1 Symptomatic Treatment

The current gold standard for treatment of AD is donepezil hydrochloride (Aricept®), a cholinesterase inhibitor, which works by inhibiting the breakdown of acetylcholine in the brain. The rationale for this drug class is to compensate for the cholinergic deficits that are consistently reported in AD by increasing cholinergic neurotransmission (Bartus 2000; Bartus et al. 1982). Two cholinesterase inhibitors have been evaluated in the canine model: the first was the gold standard, donepezil, and the second was phenserine tartrate (Araujo et al. 2011a). Donepezil (1.5 mg/kg, PO) improved memory performance on the DNMP at the longer delays during treatment compared to both baseline and washout. By contrast, phenserine (0.5 mg/kg, PO) significantly improved DNMP performance at the longest delay compared to washout, and improved learning on a difficult version of the oddity discrimination task compared to placebo. In a study replicating the effects of donepezil in which the pharmacokinetic profile was also examined, DNMP improvement corresponded with plasma levels of donepezil consistent with those reported in humans (Araujo et al. 2013b; Matsui et al. 1999). Similarly, the canine model has accurately predicted symptomatic treatments that failed human clinical trials (i.e., false positives), such as the ampakines (Studzinski et al. 2005). Collectively, these data support the predictive validity of the aged dog model for future screening of symptomatic therapeutics for AD, as well as for investigating the links among cholinergic function, Aβ pathology, and cognitive decline in both AD and CDS.

4.4.2 Disease-Modifying Treatments

The prevailing theory in AD is that Aβ is toxic and that its accumulation in the brain is one of the leading causes of cognitive dysfunction (Hardy 2006). Consequently, eliminating the accumulation of Aβ in the brain has been a major research and drug development focus since the development of the amyloid cascade hypothesis. For example, amyloid vaccines resulted in plaque clearance as well as cognitive benefits in several transgenic mouse lines (Janus and Westaway 2001; Schenk et al. 1999; Sigurdsson et al. 2001); however, the vaccination approach has repeatedly failed in phase 3 human clinical trials (Alves et al. 2014; Holmes et al. 2008).

In contrast to transgenic mouse models, the aged dog predicted the clinical outcome of active vaccine trials, thereby identifying a false-positive finding in transgenic mice (Head et al. 2008). In both old and young dogs, immunization with fibrillar $A\beta_{42}$ and a Th1 adjuvant (TiterMax Gold) resulted in primarily IgG2 and IgM antibody responses, and also caused a nonsignificant increase in CSF $A\beta_{40}$ and decrease in cortical $A\beta_{40/42}$, which was consistent with findings in transgenic models and humans (Head et al. 2006). In a subsequent 2.4-year vaccination study investigating both Aβ pathology and effect on cognitive dysfunction in aged dogs, no improvement on measures of learning, spatial attention, or spatial memory was found; however, after extended treatment, maintenance of prefrontal-dependent reversal learning ability was found (Head et al. 2008). Moreover, levels of soluble and insoluble $A\beta_{40}$ and $A\beta_{42}$ and the extent of diffuse plaque accumulation in the brain were significantly decreased in several cortical regions, with largest reductions in the prefrontal cortex. This predicted the outcome of similar human clinical vaccination trials and also suggests that reducing total Aβ may be of limited therapeutic benefit, particularly late in AD progression (Head et al. 2008; Holmes et al. 2008).

Current research extending beyond vaccination strategies includes beta- and gamma-secretase inhibitors and modulators, as well as numerous approaches targeting tau pathology. While the dog has been utilized to evaluate biomarker changes (i.e., CSF amyloid markers) following administration of secretase modulators (Borghys et al. 2012), the absence of canine studies aimed at predicting clinical trials, as well as the lack of a gold standard, limits the ability to further evaluate predictive validity of the canine model for accurately identifying successful (true-positive) disease-modifying drugs.

4.4.3 Non-pharmacological Studies

Non-pharmacological intervention studies in humans have focused primarily on lifestyle changes to see if these may improve or ameliorate the clinical signs associated with the early stages of Alzheimer's disease. Two lines of research have shown efficacious results in humans: moderate exercise and dietary modification. In humans, moderate exercise has been shown to promote brain health via changes in specific brain mechanisms (Dao et al. 2013; Radák et al. 2001; van Praag et al. 1999; Liu et al. 2011). These changes have been shown to delay the onset of cognitive delay associated with AD (e.g., Rolland et al. 2007). Aged dogs trained to run on a treadmill for 10 min daily were tested on three different memory tasks: a concurrent discrimination task, a novel object recognition task, and a novel object location task (Snigdha et al. 2014). Acute exercise was able to improve performance on both the concurrent discrimination task and the novel object location task on a 24-h retest, and chronic exercise improved performance on the object location memory task. Interestingly, one of the brain changes reported after exercise in humans is a reduction of reactive oxygen species (ROS). Alternate methods to reduce ROS are via consumption of a diet rich with antioxidants. In a large prospective study held in the Netherlands called the Rotterdam Study, it was reported that a diet high in antioxidants vitamins C and E may lower the risk for Alzheimer's disease (Engelhart et al. 2002). By comparison, a diet enriched with antioxidants alone and when combined with environmental enrichment improves cognitive function and prevents cognitive decline in aged dogs (Cotman et al. 2002; Milgram et al. 2002a, b; 2004). It is important to note that the general absence of phase 3 clinical trial data for non-pharmacological interventions impacts the ability to robustly evaluate predictive validity.

4.5 Summary

Aging dogs show AD-like cognitive domain-specific deficits and associated neuropathology that result in biomarker patterns consistent with those seen in AD progression. Therefore, the dog provides a natural model of AD progression that can be used to evaluate novel therapeutics or potentially be used to learn more about human disease progression and the sequential pathological events associated with disease progress. Moreover, aged dogs demonstrate the ability to accurately predict clinical outcomes and have successfully predicted symptomatic drugs that were approved and disease-modifying drugs that failed, such as active vaccine approaches. Consequently, the aged dog is a unique model of sporadic AD in which the progression of disease can be monitored using CSF and in vivo imaging biomarkers currently used in AD clinical research and drug development. On the other hand, developments in the AD field may also prove valuable for improved diagnosis, monitoring, and treatment of CDS.

References

Adalsteinsson E, Sullivan EV, Kleinhans N, Spielman DM, Pfefferbaum A (2000) Longitudinal decline of the neuronal marker N-acetyl aspartate in Alzheimer's disease. Lancet. 355(9216):1696–1697

Adams B, Chan A, Callahan H, Siwak C, Tapp D, Ikeda-Douglas C, Atkinson P, Head E, Cotman CW, Milgram NW (2000) Use of a delayed non-matching to position task to model age-dependent cognitive decline in the dog. Behav Brain Res 108(1):47–56

Albert MS (2011) Changes in cognition. Neurobiol Aging 32(1):S58–S63

Albert MS, DeKosky ST, Dickson D, Dubois B, Feldman HH, Fox NC, Gamst A, Holtzman DM, Jagust WJ, Petersen RC, Snyder PJ, Carrillo MC, Thies B, Phelps CH (2011) The diagnosis of mild cognitive impairment due to Alzheimer's disease: recommendations from the National Institute of Aging-Alzheimer's Association workgroups on diagnostic guidelines for Alzheimer's disease. Alzheimer's Dementia 7(3):270–279

Alves RPS, Yang MJ, Batista MT, Ferreira LCS (2014) Alzheimer's disease: is a vaccine possible? Brazil J Med Biol Res 47(6):438–444

Alzheimer A (1906) Über einen eigenartigen schweren Erkrankungsprozeß der Hirnrinde. Neurologisches Centralblatt 23:1129–1136

Amieva H, Phillips LH, Della Sala S, Henry JD (2004) Inhibitory functioning in Alzheimer's disease. Brain 127(5):949–964
Araujo JA, Chan ADF, Winka LL, Seymour PA, Milgram NW (2004) Dose-specific effects of scopolamine on canine cognition: impairment of visuospatial memory, but not visuospatial discrimination. Psychopharmacology 175(1):92–98
Araujo JA, de Rivera C, Milgram NW (2008) Visual attention is decreased in aged dogs. Alzheimer's Dementia 4(4):T205–T206
Araujo JA, Greig NH, Ingram DK, Sandin J, de Rivera C, Milgram NW (2011a) Cholinesterase inhibitors improve both memory and complex learning in aged beagle dogs. J Alzheimer's Dis 26(1):143–155
Araujo JA, Nobrega JN, Raymond R, Milgram NW (2011b) Aged dogs demonstrate both increased sensitivity to scopolamine impairment and decreased muscarinic receptor density. Pharmacol Biochem Behav 98(2):203–209
Araujo J, de Rivera C, Baulk J, Kelly S, Chakravarthy B (2013a) The reliability of and age effects on CSF measures of beta-amyloid 42 in beagle dogs: implications for a natural animal model of Alzheimer's disease progression. Alzheimer's Dementia 9(4):P851
Araujo JA, de Rivera C, Milgram NW, Sandin J (2013b) Pharmacological validation of the canine model of Alzheimer's disease: donepezil improves memory in cognitively impaired aged beagle dogs. J Alzheimer's Dis 5(4):e23
Araujo J, Sokolnicki K, Hesterman J, Hoppin J, Araujo D, Dobson H (2013c) Aged beagle dogs demonstrate reduced brain metabolism measured using PET-MR. Alzheimer's Dementia 9(4):P851
Attems J (2005) Sporadic cerebral amyloid angiopathy; pathology, clinical implications, and possible pathomechanisms. Acta Neuropathol 110(4):345–359
Attems J, Jellinger KA, Lintner F (2005) Alzheimer's disease pathology influences severity and topographical distribution of cerebral amyloid angiopathy. Acta Neuropathol 110(3):222–231
Backman L, Ginovart N, Dixon RA, Robins Wahlin TB, Wahlin A, Halldin C, Farde L (2000) Age-related cognitive deficits mediated by changes in the striatal dopamine system. Am J Psychiatry 157(4):635–637
Bartus RT (2000) On neurodegenerative diseases, models, and treatment strategies: lessons learned and lessons forgotten a generation following the cholinergic hypothesis. Exp Neurol 163(2):495–529
Bartus RT, Dean R, Beer B, Lippa AS (1982) The cholinergic hypothesis of geriatric memory dysfunction. Science 217(4558):408–414
Becker RE, Greig NH (2012) Increasing the success rate of Alzheimer's disease drug discovery and development. Expert Opin Drug Discov 7(4):367–370
Becker JT, Huff J, Nebes RD, Holland A, Boller F (1988) Neuropsychological function in Alzheimer's disease: pattern of impairment and rates of progression. Archiv Neurol 45(3):263–268
Birken DL, Oldendorf WH (1989) N-acetyl-L-aspartic acid: a literature review of a compound prominent in 1H-NMR spectroscopic studies of brain. Neurosci Biobehav Rev 13(1):23–31
Bobinski M, Wegiel J, Tarnawski M, Bobinski M, Reisberg B, De Leon MJ, Miller DC, Wisniewski HM (1997) Relationships between regional neuronal loss and neurofibrillary changes in the hippocampal formation and duration and severity of Alzheimer disease. J Neuropathol Exp Neurol 56(4):414–420
Bobinski M, De Leon MJ, Wegiel J, Desanti S, Convit A, Saint Louis LA, Rusinek H, Wisniewski HM (1999) The histological validation of post mortem magnetic resonance imaging-determined hippocampal volume in Alzheimer's disease. Neuroscience 95(3):721–725
Borghys H, Tuefferd M, Van Broeck B, Clessens E, Dillen L, Cools W, Vinken P, Straetemans R, De Ridder F, Gijsen H, Mercken M (2012) A canine model to evaluate efficacy and safety of γ-secretase inhibitors and modulators. J Alzheimer's Dis 28(4):809–822
Braak H, Braak E (1991) Neuropathological stageing of Alzheimer-related changes. Acta Neuropathol 82(4):239–259

Braak H, Braak E, Bohl J (1993) Staging of Alzheimer-related cortical destruction. Eur Neurol 33(6):403–408

Buerger K, Ewers M, Pirttilä T, Zinkowski R, Alafuzoff I, Teipel SJ, DeBernardis J, Kerkman D, McCulloch C, Soininen H, Hampel H (2006) CSF phosphorylated tau protein correlates with neocortical neurofibrillary pathology in Alzheimer's disease. Brain 129(11):3035–3041

Chan AD, Nippak P, Murphey H, Ikeda-Douglas CJ, Muggenburg B, Head E, Cotman CW, Milgram NW (2002) Visuospatial impairments in aged canines (Canis familiaris): the role of cognitive-behavioral flexibility. Behav Neurosci 116(3):443

Colle MA, Hauw JJ, Crespeau F, Uchihara T, Akiyama H, Checler F, Pageat P, Duykaerts C (2000) Vascular and parenchymal Aβ deposition in the aging dog: correlation with behavior. Neurobiol Aging 21(5):695–704

Cotman CW, Head E, Muggenburg BA, Zicker S, Milgram NW (2002) Brain aging in the canine: a diet enriched in antioxidants reduces cognitive dysfunction. Neurobiol Aging 23(5):809–818

Cummings JL (2000) Cognitive and behavioral heterogeneity in Alzheimer's disease: seeking the neurobiological basis. Neurobiol Aging. 21(6):845–861

Cummings BJ, Su JH, Cotman CW, White R, Russell MJ (1993) β-Amyloid accumulation in aged canine brain: a model of early plaque formation in Alzheimer's disease. Neurobiol Aging 14(6):547–560

Cummings BJ, Head E, Afagh AJ, Milgram NW, Cotman CW (1996a) β-amyloid accumulation correlates with cognitive dysfunction in the aged canine. Neurobiol Learning Memory 66(1):11–23

Cummings BJ, Head E, Ruehl W, Milgram NW, Cotman CW (1996b) The canine as an animal model of human aging and dementia. Neurobiol Aging 17(2):259–268

Cummings BJ, Pike CJ, Shankle R, Cotman CW (1996c) Beta-amyloid deposition and other measures of neuropathology predict cognitive status in Alzheimer's disease. Neurobiol Aging 17(6):921–933

Cummings JL, Morstorf T, Zhong K (2014) Alzheimer's disease drug development pipeline: few candidates, frequent failures. Alzheimer's Res Therapy. 6(4):37

Czasch S, Paul S, Baumgartner W (2006) A comparison of immunohistochemical and silver staining methods for the detection of diffuse plaques in the aged canine brain. Neurobiol Aging 27(2):293–305

Dao AT, Zagaar MA, Levine AT, Salim S, Eriksen JL, Alkadhi KA (2013) Treadmill exercise prevents learning and memory impairment in Alzheimer's disease-like pathology. Curr Alzheimer Res 10(5):507–515

De Leon MJ, Convit A, Wolf OT, Tarshish CY, DeSanti S, Rusinek H, Tsui W, Kandil E, Scherer AJ, Roche A, Imossi A (2001) Prediction of cognitive decline in normal elderly subjects with 2-[18F] fluoro-2-deoxy-D-glucose/positron-emission tomography (FDG/PET). Proc Natl Acad Sci 98(19):10966–10971

de Rivera C, Konyer B, Dobson H, Araujo JA, Milgram NW (2007) Magnetic resonance spectroscopy reveals similarities in brain metabolites between puppies and kittens. Society Neurosci 868.18/B12

Delano-Wood L, Bondi MW, Sacco J, Abeles N, Jak AJ, Libon DJ, Bozoki A (2009) Heterogeneity in mild cognitive impairment: differences in neuropsychological profile and associated white matter lesion pathology. J Int Neuropsychol Soc 15(6):906–914

Dimakopoulous AC, Mayer RJ (2002) Aspects of neurodegeneration in the canine brain. J Nutrition 132(6):1579S–1582S

Double KL, Halliday GM, Krill JJ, Harasty JA, Cullen K, Brooks WS, Creasey H, Broe GA (1996) Topography of brain atrophy during normal aging and Alzheimer's disease. Neurobiol Aging 17(4):513–521

Drachman DA, Leavitt J (1974) Human memory and the cholinergic system: a relationship to aging? Archives Neurol 30(2):113–121

Dringenberg HC (2000) Alzheimer's disease: more than a 'cholinergic disorder'—evidence that cholinergic–monoaminergic interactions contribute to EEG slowing and dementia. Behav Brain Res 115(2):235–249

Engelhart MJ, Geerlings MI, Ruitenberg A, van Swieten JC, Hofman A, Witteman JC, Breteler MM (2002) Dietary intake of antioxidants and risk of Alzheimer disease. J Am Med Assoc 287(24):3223–3229

Eppig J, Wambach D, Nieves C, Price CC, Lamar M, Delano-Wood L, Giovannetti T, Bettcher BM, Penney DL, Swenson R, Lippa C, Kabasakalian A, Bondi MW, Libon DJ (2012) Dysexecutive functioning in mild cognitive impairment: derailment in temporal gradients. J Int Neuropsychol Soc 18(1):20–28

Fagan AM, Mintun MA, Mach RH, Lee SY, Dence CS, Shah AR, LaRossa GN, Spinner ML, Klunk WE, Mathis CA, DeKosky ST (2006) Inverse relation between in vivo amyloid imaging load and cerebrospinal fluid Aβ42 in humans. Ann Neurol 59(3):512–519

Foldi NS, Lobosco JJ, Schaefer LA (2002) The effect of attentional dysfunction in Alzheimer's disease: theoretical and practical implications. Semin Speech Lang 23(2):139–150. Copyright © 2002 by Thieme Medical Publishers, Inc., 333 Seventh Avenue, New York, NY 10001, USA

Fox NC, Freeborough PA (1997) Brain atrophy progression measured from registered serial MRI: validation and application to Alzheimer's disease. J Magnet Reson Imag 7(6):1069–1075

Frisoni GB, Fox NC, Jack CR, Scheltens P, Thompson PM (2010) The clinical use of structural MRI in Alzheimer disease. Nat Rev Neurol 6(2):67–77

Giaccone G, Verga L, Finazzi M, Pollo B, Tagliavini F, Frangione B, Bugiani O (1990) Cerebral preamyloid deposits and congophilic angiopathy in aged dogs. Neurosci Lett 114(2):178–183

Gonzalez-Soriano J, Garcia PM, Contreras-Rodriguez J, Martinez-Sainz P, Rodriguez-Veiga E (2001) Age-related changes in the ventricular system of the dog brain. Ann Anatomy-Anatomischer Anzeiger 183(3):283–291

Grady CL, Haxby JV, Horwitz B, Sundaram M, Berg G, Schapiro M, Friedland RP, Rapoport SI (1988) Longitudinal study of the early neuropsychological and cerebral metabolic changes in dementia of the Alzheimer type. J Clin Exp Neuropsychol 10(5):576–596

Greig NH, Sambamurti K, Yu QS, Brossi A, Bruinsma GB, Lahiri DK (2005) An overview of phenserine tartrate, a novel acetylcholinesterase inhibitor for the treatment of Alzheimer's disease. Curr Alzheimer Res 2(3):281–290

Grimmer T, Riemenschneider M, Förstl H, Henriksen G, Klunk WE, Mathis CA, Shiga T, Wester HJ, Kurz A, Drzezga A (2009) Beta amyloid in Alzheimer's disease: increased deposition in brain is reflected in reduced concentration in cerebrospinal fluid. Biol Psychiatry 65(11):927–934

Grober E, Dickson D, Sliwinski MJ, Buschke H, Katz M, Crystal H, Lipton RB (1999) Memory and mental status correlates of modified Braak staging. Neurobiol Aging. 20(6):573–579

Grossman H, Bergmann C, Parker S (2006) Dementia: a brief review. Mount Sinai J Med 73(7):985–992

Grudzien A, Shaw P, Weintraub S, Bigio E, Mash DC, Mesulam MM (2007) Locus coeruleus neurofibrillary degeneration in aging, mild cognitive impairment and early Alzheimer's disease. Neurobiol Aging. 28(3):327–335

Hampel H, Bürger K, Teipel SJ, Bokde AL, Zetterberg H, Blennow K (2008) Core candidate neurochemical and imaging biomarkers of Alzheimer's disease. Alzheimer's Dementia 4(1):38–48

Hardy J (2006) Alzheimer's disease: the amyloid cascade hypothesis: an update and reappraisal. J Alzheimer's Dis 9(3):151–153

Head E, Callahan H, Muggenburg BA, Cotman CW, Milgram NW (1998) Visual-discrimination learning ability and β-amyloid accumulation in the dog. Neurobiol Aging 19(5):415–425

Head E, McCleary R, Hahn FF, Milgram NW, Cotman CW (2000) Region-specific age at onset of β-amyloid in dogs. Neurobiol Aging. 21(1):89–96

Head E, Liu J, Hagen TM, Muggenburg BA, Milgram NW, Ames BN, Cotman CW (2002) Oxidative damage increases with age in a canine model of human brain aging. J Neurochem 82(2):375–381

Head E, Barrett EG, Murphy MP, Das P, Nistor M, Sarsoza F, Glabe CC, Kayed R, Milton S, Vasilevko V, Milgram NW (2006) Immunization with fibrillar Aβ 1–42 in young and aged canines: Antibody generation and characteristics, and effects on CSF and brain Aβ. Vaccine 24(15):2824–2834

Head E, Pop V, Vasilevko V, Hill M, Saing T, Sarsoza F, Nistor M, Christie LA, Milton S, Glabe C, Barrett E (2008) A two-year study with fibrillar β-amyloid (Aβ) immunization in aged canines: effects on cognitive function and brain Aβ. J Neurosci 28(14):3555–3566

Head E, Nukala VN, Fenoglio KA, Muggenburg BA, Cotman CW, Sullivan PG (2009) Effects of age, dietary, and behavioral enrichment on brain mitochondria in a canine model of human aging. Exp Neurol 220(1):171–176

Herzig MC, Van Nostrand WE, Jucker M (2006) Mechanism of cerebral beta-amyloid angiopathy: murine and cellular models. Brain Pathol 16(1):40–54

Holmes C, Boche D, Wilkinson D, Yadegarfar G, Hopkins V, Bayer A, Jones RW, Bullock R, Love S, Neal JW, Zotova E (2008) Long-term effects of Aβ 42 immunisation in Alzheimer's disease: follow-up of a randomised, placebo-controlled phase I trial. Lancet 372(9634):216–223

Horwitz B, Tagamets MA, McIntosh AR (1999) Neural modeling, functional brain imaging, and cognition. Trends Cognit Sci 3(3):91–98

Hwang IK, Yoon YS, Yoo KY, Li H, Choi JH, Kim DW, Yi SS, Seong JK, Lee IS, Won MH (2008) Differences in lipid peroxidation and Cu, Zn-superoxide dismutase in the hippocampal CA1 region between adult and aged dogs. J Veterinary Med Sci 70(3):273–277

Hyman BT, Trojanowski JQ (1997) Consensus recommendations for the postmortem diagnosis of Alzheimer disease from the National Institute on Aging and the Reagan Institute Working Group on diagnostic criteria for the neuropathological assessment of Alzheimer disease. J Neuropathy Exp Neurol 56(10):1095–1097

Insua D, Suárez ML, Santamarina G, Sarasa M, Pesini P (2010) Dogs with canine counterpart of Alzheimer's disease lose noradrenergic neurons. Neurobiol Aging 31(4):625–635

Ishihara T, Gondo T, Takahashi M, Uchino F, Ikeda SI, Allsop D, Imai K (1991) Immunohistochemical and immunoelectron microscopical characterization of cerebrovascular and senile plaque amyloid in aged dogs' brains. Brain Res 548(1):196–205

Jack CR, Petersen RC, O'Brien PC, Tangalos EG (1992) MR-based hippocampal volumetry in the diagnosis of Alzheimer's disease. Neurology 42(1):183–183

Jack CR, Petersen RC, Xu YC, Waring SC, O'Brien PC, Tangalos EG, Smith GE, Ivnik RJ, Kokmen E (1997) Medial temporal atrophy on MRI in normal aging and very mild Alzheimer's disease. Neurology 49(3):786–794

Jack CR, Shiung MM, Weigand SD, O'Brien PC, Gunter JL, Boeve BF, Knopman DS, Smith GE, Ivnik RJ, Tangalos EG, Petersen RC (2005) Brain atrophy rates predict subsequent clinical conversion in normal elderly and amnestic MCI. Neurology 65(8):1227–1231

Jack CR, Bernstein MA, Fox NC, Thompson P, Alexander G, Harvey D, Borowski B, Britson PJ, Whitwell JL, Ward C, Dale AM (2008) The Alzheimer's disease neuroimaging initiative (ADNI): MRI methods. J Magnet Res Imag 27(4):685–691

Jack CR, Knopman DS, Jagust WJ, Shaw LM, Aisen PS, Weiner MW, Trojanowski JQ (2010) Hypothetical model of dynamic biomarkers of the Alzheimer's pathological cascade. Lancet Neurol 9(1):119–128

Jagust W, Gitcho A, Sun F, Kuczynski B, Mungas D, Haan M (2006) Brain imaging evidence of preclinical Alzheimer's disease in normal aging. Ann Neurol 59(4):673–681

Jagust WJ, Landau SM, Shaw LM, Trojanowski JQ, Koeppe RA, Reiman EM, Foster NL, Petersen RC, Weiner MW, Price JC, Mathis CA (2009) Relationships between biomarkers in aging and dementia. Neurology 73(15):1193–1199

Janus C, Westaway D (2001) Transgenic mouse models of Alzheimer's disease. Physiol Behav 73(5):873–886

Johnson DK, Storandt M, Morris JC, Galvin JE (2009) Longitudinal study of the transition from healthy aging to Alzheimer's disease. Archiv Neurol 66(10):1254–1259

Johnstone EM, Chaney MO, Norris FH, Pascual R, Little SP (1991) Conservation of the sequence of the Alzheimer's disease amyloid peptide in dog, polar bear and five other mammals by cross-species polymerase chain reaction analysis. Mol Brain Res 10(4):299–305

Kantarci KJCJ, Jack CR, Xu YC, Campeau NG, O'Brien PC, Smith GE, Ivnik RJ, Boeve BF, Kokmen E, Tangalos EG, Petersen RC (2000) Regional metabolic patterns in mild cognitive impairment and Alzheimer's disease a 1h mrs study. Neurology 55(2):210–217

Kiatipattanasakul W, Nakamura SI, Kuroki K, Nakayama H (1997) Immunohistochemical detection of anti-oxidative stress enzymes in the dog brain. Neuropathology 17(4):307–312

Kimotsuki T, Nagaoka T, Yasuda M, Tamahara S, Matsuki N, Ono K (2005) Changes of magnetic resonance imaging on the brain in beagle dogs with aging. J Veterinary Med Sci 67(10):961–967

Klafki HW, Staufenbiel M, Kornhuber J, Wiltfang J (2006) Therapeutic approaches to Alzheimer's disease. Brain 129(11):2840–2855

Lacor PN, Buniel MC, Chang L, Fernandez SJ, Gong Y, Viola KL, Lambert MP, Velasco PT, Bigio EH, Finch CE, Krafft GA (2004) Synaptic targeting by Alzheimer's-related amyloid β oligomers. J Neurosci 24(45):10191–10200

Lafleche G, Albert MS (1995) Executive function deficits in mild Alzheimer's disease. Neuropsychology 9(3):313–320

Landsberg GM, Nichol J, Araujo JA (2012) Cognitive dysfunction syndrome: a disease of canine and feline brain aging. Veterinary Clin N Am: Small Animal Pract 42(4):749–768

Libon DJ, Xie SX, Eppig J, Wicas G, Lamar M, Lippa C, Bettcher BM, Price CC, Giovannetti T, Swenson R, Wambach DM (2010) The heterogeneity of mild cognitive impairment: a neuropsychological analysis. J Int Neuropsychol Soc 16(1):84–93

Libon DJ, Bondi MW, Price CC, Lamar M, Eppig J, Wambach DM, Nieves C, Delano-Wood L, Giovannetti T, Lippa C, Kabasakalian A, Cosentino S, Swenson R, Penney DL (2011) Verbal serial list learning in mild cognitive impairment: a profile analysis of interference, forgetting, and errors. J Int Neuropsychol Soc 17(5):905–914

Linn RT, Wolf PA, Bachman DL, Knoefel JE, Cobb JL, Belanger AJ, Kaplan EF, D'Agostino RB (1995) The 'preclinical phase' of probable Alzheimer's disease: a 13-year prospective study of the Framingham cohort. Archiv Neurol 52(5):485–490

Liu HL, Zhao G, Cai K, Zhao HH (2011) Treadmill exercise prevents decline in spatial learning and memory in APP/PS1 transgenic mice through improvement of hippocampal long-term potentiation. Behav Brain Res 218(2):308–314

Marin DB, Green CR, Schmeidler J, Harvey PD, Lawlor BA, Ryan TM, Aryan M, Davis KL, Mohs RC (1997) Noncognitive disturbances in Alzheimer's disease: frequency, longitudinal course, and relationship to cognitive symptoms. J Am Geriatrics Soc. 45(11):1331–1338

Markesbery WR (1997) Oxidative stress hypothesis in Alzheimer's disease. Free Radical Biol Med 23(1):134–147

Markesbery WR (2010) Neuropathologic alterations in mild cognitive impairment: a review. J Alzheimer's Dis 19(1):221–228

Matsui K, Taniguchi S, Yoshimura T (1999) Correlation of the intrinsic clearance of donepezil (Aricept®) between in vivo and in vitro studies in rat, dog and human. Xenobiotica 29(11):1059–1072

McKhann G, Drachman D, Folstein M, Katzman R, Price D, Stadlan EM (1984) Clinical diagnosis of Alzheimer's disease. Neurology 34(7):939–944

McKhann GM, Knopman DS, Chertkow H, Hyman BT, Jack CR, Kawas CH, Klunk WE, Koroshetz WJ, Manly JJ, Mayeux R, Mohs RC (2011) The diagnosis of dementia due to Alzheimer's disease: recommendations from the National Institute on Aging-Alzheimer's Association workgroups on diagnostic guidelines for Alzheimer's disease. Alzheimer's Dementia 7(3):263–269

Milgram NW, Head E, Muggenburg B, Holowachuk D, Murphey H, Estrada J, Ikeda-Douglas CJ, Zicker SC, Cotman CW (2002a) Landmark discrimination learning in the dog: effects of age, an antioxidant fortified food, and cognitive strategy. Neurosci Biobehav Rev 26(6):679–695

Milgram NW, Zicker SC, Head E, Muggenburg BA, Murphey H, Ikeda-Douglas CJ, Cotman CW (2002b) Dietary enrichment counteracts age-associated cognitive dysfunction in canines. Neurobiol Aging 23(5):737–745

Milgram NW, Head E, Zicker SC, Ikeda-Douglas C, Murphey H, Muggenberg BA, Siwak CT, Tapp PD, Lowry SR, Cotman CW (2004) Long-term treatment with antioxidants and a program of

behavioral enrichment reduces age-dependent impairment in discrimination and reversal learning in beagle dogs. Exp Gerontol 39(5):753–765
Mirra SS, Heyman A, McKeel D, Sumi SM, Crain BJ, Brownlee LM, Vogel FS, Hughes JP, van Belle G, Berg L (1991) The consortium to establish a registry for Alzheimer's disease (CERAD) Part II. Standardization of the neuropathologic assessment of Alzheimer's disease. Neurology 41(4):479–486
Mongillo P, Araujo JA, Pitteri E, Carnier P, Adamelli S, Regolin L, Marinelli L (2013) Spatial reversal learning is impaired by age in pet dogs. Age 35(6):2273–2282
Morris JC, Aisen PS, Bateman RJ, Benzinger TL, Cairns NJ, Fagan AM, Ghetti B, Goate AM, Holtzman DM, Klunk WE, McDade E (2012) Developing an international network for Alzheimer's research: the Dominantly Inherited Alzheimer Network. Clin Investig 2(10):975–984
Morys J, Narkiewicz O, Maciejewska B, Wegiel J, Wisniewski HM (1994) Amyloid deposits and loss of neurones in the claustrum of the aged dog. Neuroreport 5(14):1825–1828
Moulder KL, Snider BJ, Mills SL, Buckles VD, Santacruz AM, Bateman RJ, Morris JC (2013) Dominantly Inherited Alzheimer Network: facilitating research and clinical trials. Alzheimer's Res Therapy 5(5):1
Murphy MP, Morales J, Beckett TL, Astarita G, Piomelli D, Weidner A, Studzinski CM, Dowling AL, Wang X, LeVine H III, Kryscio RJ (2010) Changes in cognition and amyloid-β processing with long term cholesterol reduction using atorvastatin in aged dogs. J Alzheimer's Dis 22(1):135–150
Norbury R, Travis MJ, Erlandsson K, Waddington W, Owens J, Pimlott S, Ell PJ, Murphy DGM (2005) In vivo imaging of muscarinic receptors in the aging female brain with (R, R)[123 I]-I-QNB and single photon emission tomography. Exp Gerontol 40(3):137–145
Okuda R, Uchida K, Tateyama S, Yamaguchi R, Nakayama H, Goto N (1994) The distribution of amyloid beta precursor protein in canine brain. Acta Neuropathol 87(2):161–167
Opii WO, Joshi G, Head E, Milgram NW, Muggenburg BA, Klein JB, Pierce WM, Cotman CW, Butterfield DA (2008) Proteomic identification of brain proteins in the canine model of human aging following a long-term treatment with antioxidants and a program of behavioral enrichment: relevance to Alzheimer's disease. Neurobiol Aging 29(1):51–70
Ott A, Breteler MM, Van Harskamp F, Claus JJ, Van Der Cammen TJ, Grobbee DE, Hofman A (1995) Prevalence of Alzheimer's disease and vascular dementia: association with education. The Rotterdam study. BMJ 310(6985):970–973
Papaioannou N, Tooten PC, van Ederen AM, Bohl JR, Rofina J, Tsangaris T, Gruys E (2001) Immunohistochemical investigation of the brain of aged dogs. I. Detection of neurofibrillary tangles and of 4-hydroxynonenal protein, an oxidative damage product, in senile plaques. Amyloid 8(1):11–21
Parasuraman R, Greenwood PM, Alexander GE (1995) Selective impairment of spatial attention during visual search in Alzheimer's disease. Neuroreport 6(14):1861–1864
Pekcec A, Baumgärtner W, Bankstahl JP, Stein VM, Potschka H (2008) Effect of aging on neurogenesis in the canine brain. Aging Cell 7(3):368–374
Perry E, Martin-Ruiz C, Lee M, Griffiths M, Johnson M, Piggott M, Haroutunian V, Buxbaum JD, Nasland J, Davis K, Gotti C, Clementi F, Tzartos S, Cohen O, Soreq H, Jaros E, Perry R, Ballard C, McKeith I, Court J (2000) Nicotinic receptor subtypes in human brain ageing, Alzheimer and Lewy body diseases. Eur J Pharmacol 393(1):215–222
Petersen RC (1998) Clinical subtypes of Alzheimer's disease. Dementia Geriatr Cognit Disorders 9(3):16–24
Petersen RC (2004) Mild cognitive impairment as a diagnostic entity. J Int Med 256(3):183–194
Petersen RC, Morris JC (2005) Mild cognitive impairment as a clinical entity and treatment target. Archiv Neurol 62(7):1160–1163
Petersen RC, Smith GE, Ivnik RJ, Kokmen E, Tangalos EG (1994) Memory function in very early Alzheimer's disease. Neurology 44(5):867–867

Petrie EC, Cross DJ, Galasko D, Schellenberg GD, Raskind MA, Peskind ER, Minoshima S (2009) Preclinical evidence of Alzheimer changes: convergent cerebrospinal fluid biomarker and fluorodeoxyglucose positron emission tomography findings. Archiv Neurol 66(5):632–637

Picciotto MR, Zoli M (2002) Nicotinic receptors in aging and dementia. J Neurobiol 53(4):641–655

Pop V, Head E, Berchtold NC, Glabe CG, Studzinski CM, Weidner AM, Murphy MP, Cotman CW (2012) Aβ aggregation profiles and shifts in APP processing favor amyloidogenesis in canines. Neurobiol Aging. 33(1):108–120

Portelius E, Van Broeck B, Andreasson U, Gustavsson MK, Mercken M, Zetterberg H, Borghys H, Blennow K (2010) Acute effect on the Aβ isoform pattern in CSF in response to γ-secretase modulator and inhibitor treatment in dogs. J Alzheimer's Dis 21(3):1005–1012

Prior R, D'Urso D, Frank R, Prikulis I, Pavlakovic G (1996) Loss of vessel wall viability in cerebral amyloid angiopathy. Neuroreport 7(2):562–564

Pugliese M, Mascort J, Mahy N, Ferrer I (2006) Diffuse beta-amyloid plaques and hyperphosphorylated tau are unrelated processes in aged dogs with behavioral deficits. Acta Neuropathol 112(2):175–183

Pugliese M, Gangitano C, Ceccariglia S, Carrasco JL, Del Fà A, Rodríguez MJ, Michetti F, Mascort J, Mahy N (2007) Canine cognitive dysfunction and the cerebellum: acetylcholinesterase reduction, neuronal and glial changes. Brain Res 1139:85–94

Qiu C, Kivipelto M, von Strauss E (2009) Epidemiology of Alzheimer's disease: occurrence, determinants, and strategies toward intervention. Dialogues Clin Neurosci 11(2):111–128

Radák Z, Kaneko T, Tahara S, Nakamoto H, Pucsok J, Sasvári M, Nyakas C, Goto S (2001) Regular exercise improves cognitive function and decreases oxidative damage in rat brain. Neurochem Int 38(1):17–23

Ramani A, Jensen JH, Helpern JA (2006) Quantitative MR Imaging in Alzheimer Disease 1. Radiology 241(1):26–44

Reid W, Broe G, Creasey H, Grayson D, McCusker E, Bennett H, Longley W, Sulway MR (1996) Age at onset and pattern of neuropsychological impairment in mild early-stage Alzheimer disease: a study of a community-based population. Archiv Neurol 53(10):1056–1061

Reinikainen KJ, Riekkinen PJ, Halonen T, Laakso M (1987) Decreased muscarinic receptor binding in cerebral cortex and hippocampus in Alzheimer's disease. Life Sci 41(4):453–461

Reinikainen KJ, Soininen H, Riekkinen PJ (1990) Neurotransmitter changes in Alzheimer's disease: implications to diagnostics and therapy. J Neurosci Res 27(4):576–586

Ritchie K, Touchon J (2000) Mild cognitive impairment: conceptual basis and current nosological status. Lancet 355(9199):225–228

Rofina J, Andel IV, Van Ederen AM, Papaioannou N, Yamaguchi H, Gruys E (2003) Canine counterpart of senile dementia of the Alzheimer type: amyloid plaques near capillaries but lack of spatial relationship with activated microglia and macrophages. Amyloid 10(2):86–96

Rofina JE, Singh K, Skoumalova-Vesela A, van Ederen AM, van Asten AJ, Wilhelm J, Gruys E (2004) Histochemical accumulation of oxidative damage products is associated with Alzheimer-like pathology in the canine. Amyloid 11(2):90–100

Rofina JE, Van Ederen AM, Toussaint MJM, Secreve M, Van Der Spek A, Van Der Meer I, Van Eerdenburg FJCM, Gruys E (2006) Cognitive disturbances in old dogs suffering from the canine counterpart of Alzheimer's disease. Brain Res 1069(1):216–226

Rolland Y, Pillard F, Klapouszczak A, Reynish E, Thomas D, Andrieu S, Rivière D, Vellas B (2007) Exercise program for nursing home residents with Alzheimer's disease: a 1-year randomized, controlled trial. J Am Geriatr Soc 55(2):158–165

Rose S, de Zubicaray G, Wang D, Galloway G, Chalk J, Eagle S, Semple J, Doddrell D (1999) A 1H MRS study of probable Alzheimer's disease and normal aging: implications for longitudinal monitoring of dementia progression. Magn Res Imag 17(2):291–299

Rossor M, Iversen LL (1986) Non-cholinergic neurotransmitter abnormalities in Alzheimer's disease. Br Med Bullet 42(1):70–74

Russell MJ, White R, Patel E, Markesbery WR, Watson CR, Geddes JW (1992) Familial influence on plaque formation in the beagle brain. Neuroreport 3(12):1093–1096

Scheltens P, Fox N, Barkhof F, De Carli C (2002) Structural magnetic resonance imaging in the practical assessment of dementia: beyond exclusion. Lancet Neurol 1(1):13–21

Schenk D, Barbour R, Dunn W, Gordon G, Grajeda H, Guido T, Hu K, Huang J, Johnson-Wood K, Khan K, Kholodenko D (1999) Immunization with amyloid-β attenuates Alzheimer-disease-like pathology in the PDAPP mouse. Nature 400(6740):173–177

Schmidt F, Boltze J, Jäger C, Hofmann S, Willems N, Seeger J, Härtig W, Stolzing A (2015) Detection and Quantification of β-Amyloid, Pyroglutamyl Aβ, and Tau in Aged Canines. J Neuropathol Exp Neurol 74(9):912–923

Selkoe DJ (1996) Amyloid β-protein and the genetics of Alzheimer's disease. J Biol Chem 271(31):18295–18298

Selkoe DJ (2001) Alzheimer's disease: genes, proteins, and therapy. Physiol Rev 81(2):741–766

Selkoe DJ (2008) Soluble oligomers of the amyloid β-protein impair synaptic plasticity and behavior. Behav Brain Res 192(1):106–113

Selkoe DJ, Bell DS, Podlisny MB, Price DL, Cork LC (1987) Conservation of brain amyloid proteins in aged mammals and humans with Alzheimer's disease. Science 235(4791):873–877

Shaw LM, Korecka M, Clark CM, Lee VMY, Trojanowski JQ (2007) Biomarkers of neurodegeneration for diagnosis and monitoring therapeutics. Nat Rev Drug Discov 6(4):295–303

Sigurdsson EM, Scholtzova H, Mehta PD, Frangione B, Wisniewski T (2001) Immunization with a nontoxic/nonfibrillar amyloid-β homologous peptide reduces Alzheimer's disease-associated pathology in transgenic mice. Am J Pathol 159(2):439–447

Simmons ML, Frondoza CG, Coyle JT (1991) Immunocytochemical localization of N-acetyl-aspartate with monoclonal antibodies. Neuroscience 45(1):37–45

Siwak CT, Tapp PD, Milgram NW (2001) Effect of age and level of cognitive function on spontaneous and exploratory behaviors in the beagle dog. Learning Memory 8(6):317–325

Siwak CT, Tapp PD, Zicker SC, Murphey HL, Muggenburg BA, Head E, Cotman CW, Milgram NW (2003) Locomotor activity rhythms in dogs vary with age and cognitive status. Behav Neurosci 117(4):813

Siwak-Tapp CT, Head E, Muggenburg BA, Milgram NW, Cotman CW (2007) Neurogenesis decreases with age in the canine hippocampus and correlates with cognitive function. Neurobiol Learn Memory 88(2):249–259

Siwak-Tapp CT, Head E, Muggenburg BA, Milgram NW, Cotman CW (2008) Region specific neuron loss in the aged canine hippocampus is reduced by enrichment. Neurobiol Aging 29(1):39–50

Skoumalova A, Rofina J, Schwippelova Z, Gruys E, Wilhelm J (2003) The role of free radicals in canine counterpart of senile dementia of the Alzheimer type. Exp Gerontol 38(6):711–719

Smith MA, Rottkamp CA, Nunomura A, Raina AK, Perry G (2000) Oxidative stress in Alzheimer's disease. Biochem Biophys Acta Mol Basis Dis 1502(1):139–144

Smolek T, Madari A, Farbakova J, Kandrac O, Jadhav S, Cente M, Brezovakova V, Zilka N (2016) Tau hyperphosphorylation in synaptosomes and neuroinflammation are associated with canine cognitive impairment. J Comparative Neurol 524(4):874–895

Snigdha S, Christie LA, De Rivera C, Araujo JA, Milgram NW, Cotman CW (2012) Age and distraction are determinants of performance on a novel visual search task in aged beagle dogs. Age 34(1):67–73

Snigdha S, de Rivera C, Milgram NW, Cotman C (2014) Exercise enhances memory consolidation in the aging brain. Front Aging Neurosci 6:3

Sperling RA, Aisen PS, Beckett LA, Bennett DA, Craft S, Fagan AM, Iwatsubo T, Jack CR, Kaye J, Montine TJ, Park DC (2011) Toward defining the preclinical stages of Alzheimer's disease: recommendations from the National Institute on Aging-Alzheimer's Association workgroups on diagnostic guidelines for Alzheimer's disease. Alzheimer's Dementia 7(3):280–292

Studzinski CM, Araujo JA, Milgram NW (2005) The canine model of human cognitive aging and dementia: pharmacological validity of the model for assessment of human cognitive-enhancing drugs. Progr Neuro-Psychopharmacol Biol Psychiatr 29(3):489–498

Studzinski CM, Christie LA, Araujo JA, Burnham WM, Head E, Cotman CW, Milgram NW (2006) Visuospatial function in the beagle dog: an early marker of cognitive decline in a model of human aging and dementia. Neurobiol Learning Memory 86(2):197–204

Su MY, Head E, Brooks WM, Wang Z, Muggenburg BA, Adam GE, Sutherland R, Cotman CW, Nalcioglu O (1998) Magnetic resonance imaging of anatomic and vascular characteristics in a canine model of human aging. Neurobiol Aging 19(5):479–485

Su MY, Tapp PD, Vu L, Chen YF, Chu Y, Muggenburg B, Chiou JY, Chen C, Wang J, Bracco C, Head E (2005) A longitudinal study of brain morphometrics using serial magnetic resonance imaging analysis in a canine model of aging. Progr Neuro-Psychopharmacol Biol Psychiatry 29(3):389–397

Sutphen CL, Jasielec MS, Shah AR, Macy EM, Xiong C, Vlassenko AG, Benzinger TL, Stoops EE, Vanderstichele HM, Brix B, Darby HD (2015) Longitudinal cerebrospinal fluid biomarker changes in preclinical Alzheimer disease during middle age. JAMA Neurol 72(9):1029–1042

Tapp PD, Siwak CT, Head E, Cotman CW, Milgram NW (2001) Sex differences in the effect of oestrogen on size discrimination learning and spatial memory. In: Proceedings of the Third International Congress on Veterinary Behavioral Medicine. Universities Federation for Animal Welfare, Wheathamstead, pp 136–138

Tapp PD, Siwak CT, Estrada J, Head E, Muggenburg BA, Cotman CW, Milgram NW (2003) Size and reversal learning in the beagle dog as a measure of executive function and inhibitory control in aging. Learning Memory. 10(1):64–73

Tapp PD, Siwak CT, Gao FQ, Chiou JY, Black SE, Head E, Muggenburg BA, Cotman CW, Milgram NW, Su MY (2004) Frontal lobe volume, function, and β-amyloid pathology in a canine model of aging. J Neurosci 24(38):8205–8213

Thal DR, Rüb U, Orantes M, Braak H (2002) Phases of Aβ-deposition in the human brain and its relevance for the development of AD. Neurology 58(12):1791–1800

Tolboom N, van der Flier WM, Yaqub M, Boellaard R, Verwey NA, Blankenstein MA, Windhorst AD, Scheltens P, Lammertsma AA, van Berckel BN (2009) Relationship of cerebrospinal fluid markers to 11C-PiB and 18F-FDDNP binding. J Nucl Med 50(9):1464–1470

Tomic JL, Pensalfini A, Head E, Glabe CG (2009) Soluble fibrillar oligomer levels are elevated in Alzheimer's disease brain and correlate with cognitive dysfunction. Neurobiol Dis 35(3):352–358

Torp R, Head E, Cotman CW (2000a) Ultrastructural analyses of β-amyloid in the aged dog brain: Neuronal β-amyloid is localized to the plasma membrane. Progr Neuro-Psychopharmacol Biol Psychiatry 24(5):801–810

Torp R, Head E, Milgram NW, Hahn F, Ottersen OP, Cotman CW (2000b) Ultrastructural evidence of fibrillar β-amyloid associated with neuronal membranes in behaviorally characterized aged dog brains. Neuroscience 96(3):495–506

Uchida K, Nakayama H, Goto N (1991) Pathological studies on cerebral amyloid angiopathy senile plaques and amyloid deposition in visceral organs in aged dogs. J Veterinary Med Sci 53(6):1037–1042

Uchida K, Tani Y, Uetsuka K, Nakayama H, Goto N (1992) Immunohistochemical studies on canine cerebral amyloid angiopathy and senile plaques. J Veterinary Med Sci 54(4):659–667

Uzun S, Kozumplik O, Folnegović-Šmalc V (2011) Alzheimer's dementia: current data review. Colleg Antropol 35(4):1333–1337

Van Praag H, Christie BR, Sejnowski TJ, Gage FH (1999) Running enhances neurogenesis, learning, and long-term potentiation in mice. Proc Natl Acad Sci 96(23):13427–13431

Vandevelde M, Higgins R, Oevermann A (2012) Veterinary neuropathology: essentials of theory and practice. Wiley, New York

Vemuri P, Wiste HJ, Weigand SD, Shaw LM, Trojanowski JQ, Weiner MW, Knopman DS, Petersen RC, Jack CR, Alzheimer's Disease Neuroimaging Initiative (2009a) MRI and CSF biomarkers in normal, MCI, and AD subjects: diagnostic discrimination and cognitive correlations. Neurology 73(4):287–293

Vemuri P, Wiste HJ, Weigand SD, Shaw LM, Trojanowski JQ, Weiner MW, Knopman DS, Petersen RC, Jack CR, Alzheimer's Disease Neuroimaging Initiative (2009b) MRI and CSF biomarkers in normal, MCI, and AD subjects: predicting future clinical change. Neurology 73(4):294–301
Vemuri P, Wiste HJ, Weigand SD, Knopman DS, Trojanowski JQ, Shaw LM, Bernstein MA, Aisen PS, Weiner M, Petersen RC, Jack CR (2010) Serial MRI and CSF biomarkers in normal aging, MCI, and AD. Neurology 75(2):143–151
Verhoeff NPLG, Wilson AA, Nobrega J, Milgram NW, Westaway D, Head E, Hussey D, Tapp D, Trop L, Giuliano F, Araujo J, Ginovart N, Richardson L, Singh K, Houle S (2003) Development of positron emission tomography (PET) tracers for beta-amyloid imaging in vivo. J Cerebral Blood Flow Metab 23(1):685
Weiner MW, Veitch DP, Aisen PS, Beckett LA, Cairns NJ, Green RC, Harvey D, Jack CR, Jagust W, Liu E, Morris JC (2013) The Alzheimer's Disease Neuroimaging Initiative: a review of papers published since its inception. Alzheimer's Dementia 9(5):111–194
West MJ (1993) Regionally specific loss of neurons in the aging human hippocampus. Neurobiol Aging 14(4):287–293
Wisniewski H, Johnson AB, Raine CS, Kay WJ, Terry RD (1970) Senile plaques and cerebral amyloidosis in aged dogs. A histochemical and ultrastructural study. Lab Investig 23:287–296
Wisniewski HM, Wegiel J, Morys J, Bancher C, Soltysiak Z, Kim KS (1990) Aged dogs: an animal model to study beta-protein amyloidogenesis. In: Alzheimer's disease. Epidemiology, neuropathology, neurochemistry, and clinics. Springer, Vienna, pp 151–168
World Health Organization (2012) Dementia: a public health priority. World Health Organization, Geneva
Yoshino T, Uchida K, Tateyama S, Yamaguchi R, Nakayama H, Goto N (1996) A retrospective study of canine senile plaques and cerebral amyloid angiopathy. Veterinary Pathol Online 33(2):230–234
Yu CH, Song GS, Yhee JY, Kim JH, Im KS, Nho WG, Lee JH, Sur JH (2011) Histopathological and immunohistochemical comparison of the brain of human patients with Alzheimer's disease and the brain of aged dogs with cognitive dysfunction. J Comparative Pathol 145(1):45–58
Zahs KR, Ashe KH (2010) 'Too much good news'–are Alzheimer mouse models trying to tell us how to prevent, not cure, Alzheimer's disease? Trends Neurosci 33(8):381–389

Brain Aging in the Dog

5

Jeff Nichol and Elizabeth Head

Dogs develop behavioral and cognitive dysfunction with age. Interestingly, as with humans, not all aged dogs become impaired, and there can be significant individual variability. Studies of the brains of cognitively characterized aged dogs suggest several possible underlying neurobiological mechanisms for observed impairments. In this chapter, changes in canine brains associated with atrophy, neuron loss, accumulation of beta-amyloid (Aβ), mitochondrial dysfunction, and resulting accumulation of oxidative damage are described. There are many important features of brain aging in dogs that overlap significantly with human brain aging, suggesting they are a useful model system in which to test interventions that may lead to healthy aging.

5.1 Introduction

Several comprehensive reviews of the neuropathology of canine brain aging are available; this chapter will serve as an overview of key features that are relevant for cognitive decline (Bosch et al. 2012; Cotman and Head 2008; Head 2000, 2011, 2013; Schutt et al. 2016). The features include brain atrophy, neuron loss, and accumulation of Alzheimer's disease (AD) like neuropathology, vascular pathology, oxidative damage, and inflammation. Most of these neuropathologies increase with age in dogs and in several studies of cognitively characterized animals are also observed to be correlated with the extent of cognitive decline.

J. Nichol
Veterinary Emergency and Specialty Centers of New Mexico, Albuquerque, NM 87109, USA

E. Head (✉)
Sanders-Brown Center on Aging, Department of Pharmacology and Nutritional Sciences, University of Kentucky, Lexington, KY 40536, USA
e-mail: Elizabeth.Head@uky.edu

G. Landsberg et al. (eds.), *Canine and Feline Dementia*,
DOI 10.1007/978-3-319-53219-6_5

5.2 Structural Brain Changes and Neuron Losses

A consistent feature of aging in humans is progressive brain atrophy, which is particularly striking in AD. Magnetic resonance imaging (MRI) studies have provided useful outcome measures reflecting changes in brain structure in vivo. Both generalized cortical atrophy (Su et al. 1998) and ventricular widening (Gonzalez-Soriano et al. 2001; Kimotsuki et al. 2005; Su et al. 1998) can be observed in older animals. However, as with various cognitive functions being differentially affected by the aging process, MRI studies suggest not all brain regions atrophy at the same rate.

Losses in tissue volume occur early with age in the prefrontal cortex around 8 – 11 years in beagles (this may differ with various breeds, given that larger breeds tend to have shorter life spans) (Tapp et al. 2004). The hippocampus atrophies in beagles after 11 years of age (Tapp et al. 2004). Paralleling correlations in cognition and frontal cortex volume in people with dementia (Du et al. 2006; Ezekiel et al. 2004), canine prefrontal cortex atrophy is also associated with impaired cognition (Rofina et al. 2006).

Changes in neuronal number or density are observed in normal human brain aging (Šimić et al. 1997; West 1993) and are more extensive in AD (Bobinski et al. 1997; West et al. 2000). In beagles, a selective loss of neurons is observed within the hilus of the hippocampus (up to 30% loss) when comparing young dogs (3 – 5 years) to older dogs (13 – 15 years) (Siwak-Tapp et al. 2008). In addition, hilar neuron number was correlated with cognition; higher numbers of neurons were associated with fewer errors on a visual discrimination task (Siwak-Tapp et al. 2008).

The reason for losses of neurons in the hippocampus may be due to slower replacement by neurogenesis. Interestingly, neurogenesis decreases by up to 90 – 95% in aging beagles (Hwang et al. 2007; Pekcec et al. 2008; Siwak-Tapp et al. 2007). Further, animals with fewer new neurons had higher error scores in measures of learning and memory, suggesting a link between neurogenesis and cognition with age (Siwak-Tapp et al. 2007). Neuron loss and cortical atrophy in vulnerable brain regions of the aged dog may be due to multiple neurodegenerative processes associated with the up- or downregulation of molecular pathways (Swanson et al. 2007). One of these pathways may be a decrease in growth factors such as brain-derived neurotrophic factor (Fahnestock et al. 2010).

5.3 Plaques and Aβ Accumulation

Plaques are extracellular deposits that contain the Aβ peptide, which is a 40–42 amino-acid-long cleavage product of the longer amyloid precursor protein (Selkoe 1994). Aβ accumulation can occur in different types of plaques (e.g., diffuse, neuritic), but also can form structures called oligomers, which are particularly toxic to synapses (Haass and Selkoe 2007; Lesné et al. 2006; Selkoe 2008; Walsh et al. 2002). A useful feature of the aged canine model is that the Aβ sequence is identical to that of humans (Johnstone et al. 1991; Selkoe et al. 1987). Indeed, it was this finding that led to interest in studying aging dogs as an approach to understand human aging and AD (Wisniewski et al. 1990).

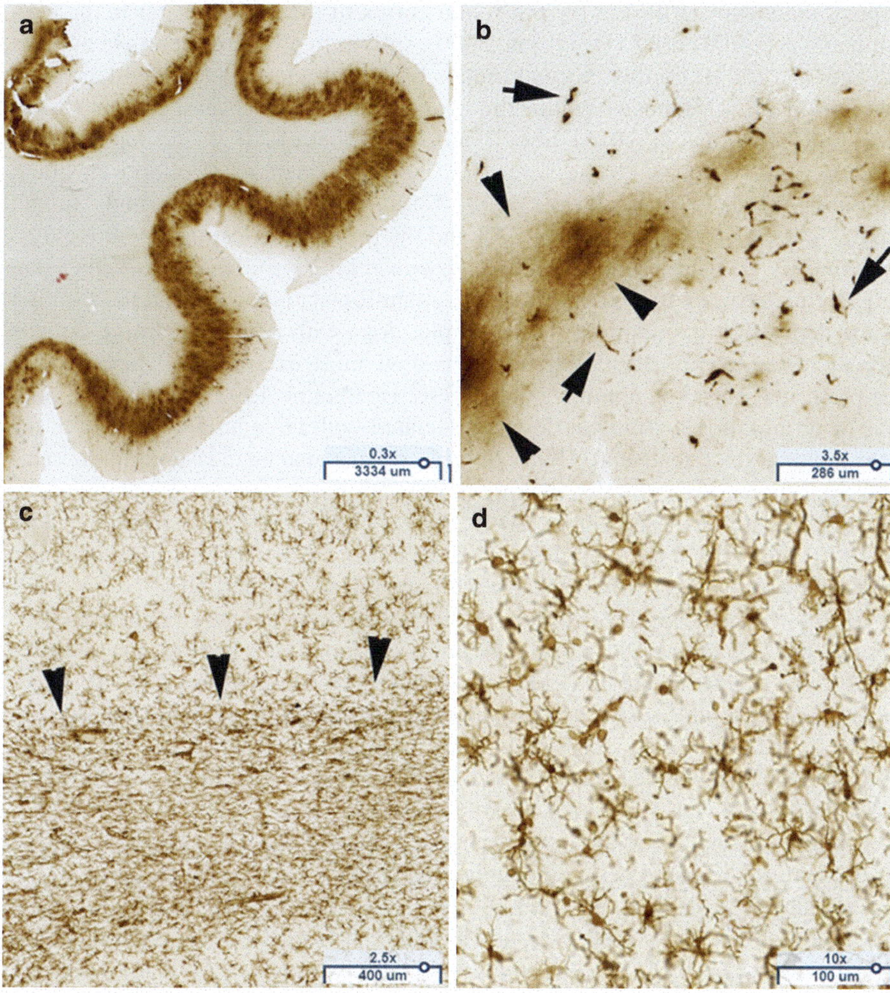

Fig. 5.1 Neuropathology in aging dogs. (**a**) Aβ immunostaining (6E10 antibody) in the parietal cortex of an aged pet dog (15-year-old Siberian Husky) shows extensive plaque deposition affecting deep cortical layers. (**b**) CAA (6E10 immunostaining—*arrows*) in the parietal cortex of an aged dog (14-year-old Sheltie) shows that vascular pathology can be extensive and tends to occur in clusters. Diffuse plaques are also identified by arrowheads. (**c**) Low-power photograph showing extensive microglial cell labeling (IBA-1 antibody) in gray matter along with intense labeling in the white matter (area below *arrows*) of a 15-year-old Shih Tzu. (**d**) Higher magnification photograph from (**c**) showing that individual microglia contain phagocytic vacuoles and have thickened processes suggesting that some have an activated morphology

In dogs, as shown in Fig. 5.1, Aβ accumulates in the cortex with a relatively well-described pattern that parallels observations in the human brain (Braak and Braak 1991; Braak et al. 1993; Giaccone et al. 1990; Head et al. 2000; Ishihara et al. 1991; Selkoe et al. 1987; Thal et al. 2002; Wisniewski et al. 1970, 1990). Aβ deposition occurs earliest in the prefrontal cortex of the dog and later in the temporal and occipital cortex (Head et al. 2000), similar to previous reports in humans (Thal et al., 2002). Further, several groups have characterized the age-dependent maturation of Aβ deposits within the canine cortex into several phases (Satou et al. 1997; Schutt et al. 2016). Importantly, the extent of Aβ plaque deposition in the dog brain is linked to the severity of cognitive deficits (Colle et al. 2000; Cummings et al. 1996; Head et al. 1998; Rofina et al. 2006). Interestingly, not all studies show a correlation between Aβ and the presence of canine cognitive dysfunction (CCD) (Chambers et al. 2011; Ozawa et al. 2016). However, studies that show a link between the extent of Aβ and cognition also indicate that the location of the deposition is important. For example, dogs with reversal learning deficits indicative of executive dysfunction tend to show more extensive Aβ deposition in the prefrontal cortex (Cummings et al. 1996; Head et al. 1998). In contrast, poor size discrimination learning ability is associated with large amounts of Aβ in the entorhinal cortex (Head et al. 1998). Soluble Aβ can also be measured in the cerebrospinal fluid (CSF) of dogs, making it a useful marker for aging and cognition intervention studies (Head et al. 2010; Sarasa et al. 2013). The ratio of Aβ42/Aβ40 in the CSF is a good predictor of the extent of Aβ measured biochemically in the brain and also declines linearly with age (Head et al. 2010).

In studies characterizing Aβ deposits in plaques, the primary conformation is fibrillar, but Aβ not only exists in this or linear conformations but can also assemble into soluble states that are toxic to synapses and thus neuronal function. Aβ oligomers are small soluble assembly states that interfere with synaptic function and cognition (Kayed et al. 2003; Walsh et al. 2002). Interestingly, Aβ oligomers in the CSF of dogs are inversely related to the amount of total Aβ measured biochemically in the brain, suggesting that oligomers are sequestered into plaques (Head et al. 2010).

Posttranslationally modified Aβ has also been assessed in aging dog brains. These modified forms of Aβ are thought to represent deposits of Aβ that are chronobiologically older (Azizeh et al. 2000). Pyroglutamyl-modified Aβ increases with age in dogs (Frost et al. 2013; Schutt et al. 2016) as does the amount of racemized Aβ (Azizeh et al. 2000). In addition to modified Aβ, truncated Aβ that is observed in AD brain is also observed in the aged canine brain (Chambers et al. 2011).

5.4 Vascular Neuropathology

A frequent pathology detected in both normal human brain aging and particularly in AD is the presence of cerebral amyloid angiopathy (CAA), which is described as the accumulation of Aβ in the walls of cerebral vessels (Attems 2005; Attems et al. 2005; Herzig et al. 2006). Aged dogs are also vulnerable to vascular pathology with perivascular abnormalities; CAA is observed frequently in aged dogs (Giaccone et al. 1990; Ishihara et al. 1991; Shimada et al. 1992; Uchida et al. 1990, 1991, 1992, 1993, 1997; Yoshino et al. 1996). The consequences of CAA accumulating in the brains of aging dogs may be a compromise to the function of the blood-brain barrier and impaired vascular function (Prior et al. 1996). In turn, vascular dysfunction and BBB disruptions may lead to microhemorrhages (Deane and Zlokovic 2007; Uchida et al. 1990, 1991). The distribution of CAA in dog brain is similar to humans, with particular vulnerability in the occipital cortex (Attems et al. 2005). However, in a systematic study of the extent of CAA in cognitively characterized pet dogs, CAA increased with age but did not correlate with cognition (Ozawa et al. 2016). Thus, aged dogs develop cerebrovascular abnormalities that may not contribute to cognitive decline but are otherwise consistent with those reported in humans.

5.5 Oxidative Damage and Mitochondrial Dysfunction

There are several reviews of the potential role for oxidative damage and mitochondrial dysfunction on brain aging in dogs (Cotman et al. 2002; Dowling and Head 2012). The production of free radicals during the aging process can lead to damaged proteins, lipids, and nucleotides, which may cause neuronal dysfunction and degeneration. The aging dog brain accumulates carbonyl groups, a measure of oxidative damage to proteins (Head et al. 2002; Skoumalova et al. 2003). Typically, the activity of endogenous antioxidants balances the production of free radicals. However, several of these protective mechanisms decline with age. Carbonyl groups are associated with reduced endogenous antioxidant enzyme activity/protein levels, including those of glutamine synthetase and superoxide dismutase (SOD) (Head et al. 2002; Hwang et al. 2008; Kiatipattanasakul et al. 1997; Opii et al. 2008). The end products of lipid peroxidation (oxidative damage to lipids), including 4-hydroxynonenal (4HNE) (Hwang et al. 2008; Papaioannou et al. 2001; Rofina et al. 2004, 2006), lipofuscin (Rofina et al. 2006), lipofuscin-like pigments (Papaioannou et al. 2001; Rofina et al. 2004), or malondialdehyde (Head et al. 2002); all of these increase with age in the canine brain. In addition, there are several reports of increased oxidative damage to DNA or RNA in the aged dog brain (Cotman and Head 2008; Rofina et al. 2006).

The consequences of increasing oxidative damage with age in dogs may be compromised neuronal function leading to deficits in cognition. In aged pet dogs, higher levels of oxidative end products are correlated with more severe behavioral changes (Rofina et al. 2004, 2006; Skoumalova et al. 2003). Similarly, in studies of laboratory beagles, higher protein oxidative damage (3-nitrotyrosine) and lower endogenous antioxidant capacity (SOD and glutathione-S-transferase activities) in aging animals are associated with poorer prefrontal-dependent learning and spatial learning (Opii et al. 2008). The mitochondria are a critical contributor to oxidative damage, being a source of free radicals that damage proteins, lipids, and DNA/RNA (Shigenaga et al. 1994). Mitochondria isolated from young beagle brain have lower levels of reactive oxygen species production than those isolated from older beagles (Head et al. 2009). Due to the similarities in oxidative damage in dogs and humans, several research groups have suggested that the canine is particularly well suited for studies focused on this mechanism of neurodegeneration (Dowling and Head 2012; Romanucci and Della Salda 2015).

5.6 Inflammation

Neuroinflammation in the aged human and AD brain may lead to the exacerbation of cognitive decline or potentially mediate other neuropathological events causing dementia (Heneka et al. 2015; Wilcock 2013). Although not as well characterized as neuroinflammation in the human brain, there are several small studies in aged pet dog brains. In a recent study, Schutt and colleagues (Schutt et al. 2016) measured canine cytokines in the prefrontal cortex of 15 aged dogs as compared with 2 young dogs. Pro-inflammatory cytokines were generally at low levels and were not associated with the extent of cognitive dysfunction. However, using measures of glial activation (microglial cells and astrocytes), increasing numbers of both types of cells were associated with more extensive CCD in a study of 37 dogs with various breeds included (Ozawa et al. 2016). Similarly, the level of S100β astrocytosis, a putative measure of inflammation, is also correlated with cognitive deficits in pet dogs (Pugliese et al. 2006).

5.7 White Matter Pathology

White matter degeneration can contribute to cognitive decline in humans (Bartzokis 2004; Gold et al. 2012). In a study of myelin protein levels as a function of age in dogs, Chambers et al. report a loss of prefrontal myelin protein (Chambers et al. 2012) that was also associated with some Aβ deposition in CAA. In another study, the extent of ubiquitin labeling in aging dog brains ($n = 37$) was associated with CCD and is thought to reflect failures in the protein homeostasis in the synapse or in myelin (Ozawa et al. 2016).

5.8 Summary

Identifying neurodegenerative mechanisms underlying cognitive dysfunction in aging dogs will provide novel targets for future intervention studies. These treatments may involve lifestyle changes (e.g., exercise can lead to enhanced neurogenesis and benefit cognition in dogs (Intlekofer and Cotman 2013; Snigdha et al. 2014)), diet changes (e.g., antioxidant diets (Cotman et al. 2002)), removal of Aβ by vaccines (Cotman and Head 2008), or pharmacological manipulations. As more of the gaps in our knowledge are filled, we may learn more of the role of vascular factors and inflammation in canine brain aging, which are also modifiable and may lead to cognitive benefits.

Acknowledgment Funding provided by the NIH/NIA R01AG0031764.

References

Attems J (2005) Sporadic cerebral amyloid angiopathy: pathology, clinical implications, and possible pathomechanisms. Acta Neuropathol 110(4):345–359

Attems J, Jellinger KA, Lintner F (2005) Alzheimer's disease pathology influences severity and topographical distribution of cerebral amyloid angiopathy. Acta Neuropathol 110(3):222–231

Azizeh BY, Head E, Ibrahim MA, Torp R, Tenner AJ, Kim RC, Lott IT, Cotman CW (2000) Molecular dating of senile plaques in the brains of individuals with Down syndrome and in aged dogs. Exp Neurol 163(1):111–122

Bartzokis G (2004) Age-related myelin breakdown: a developmental model of cognitive decline and Alzheimer's disease. Neurobiol Aging 25(1):5–18. author reply 49–62

Bobinski M, Wegiel J, Tarnawski M, Bobinski M, Reisberg B, de Leon MJ, Miller DC, Wisniewski HM (1997) Relationships between regional neuronal loss and neurofibrillary changes in the hippocampal formation and duration and severity of Alzheimer disease. J Neuropathol Exp Neurol 56(4):414–420

Bosch MN, Pugliese M, Gimeno-Bayon J, Rodriguez MJ, Mahy N (2012) Dogs with cognitive dysfunction syndrome: a natural model of Alzheimer's disease. Curr Alzheimer Res 9(3):298–314

Braak H, Braak E (1991) Neuropathological stageing of Alzheimer-related changes. Acta Neuropathol 82(4):239–259

Braak H, Braak E, Bohl J (1993) Staging of Alzheimer-related cortical destruction. Eur Neurol 33(6):403–408

Chambers JK, Mutsuga M, Uchida K, Nakayama H (2011) Characterization of AβpN3 deposition in the brains of dogs of various ages and other animal species. Amyloid 18(2):63–71

Chambers JK, Uchida K, Nakayama H (2012) White matter myelin loss in the brains of aged dogs. Exp Gerontol 47(3):263–269

Colle M-A, Hauw J-J, Crespeau F, Uchiara T, Akiyama H, Checler F, Pageat P, Duykaerts C (2000) Vascular and parenchymal Aβ deposition in the aging dog: correlation with behavior. Neurobiol Aging 21(5):695–704

Cotman CW, Head E (2008) The canine (dog) model of human aging and disease: dietary, environmental and immunotherapy approaches. J Alzheimer's Dis 15(4):685–707

Cotman CW, Head E, Muggenburg BA, Zicker S, Milgram NW (2002) Brain aging in the canine: a diet enriched in antioxidants reduces cognitive dysfunction. Neurobiol Aging 23(5):809–818

Cummings BJ, Head E, Afagh AJ, Milgram NW, Cotman CW (1996) Beta-amyloid accumulation correlates with cognitive dysfunction in the aged canine. Neurobiol Learning Memory 66(1):11–23

Deane R, Zlokovic BV (2007) Role of the blood-brain barrier in the pathogenesis of Alzheimer's disease. Curr Alzheimer Res 4(2):191–197

Dowling AL, Head E (2012) Antioxidants in the canine model of human aging. Biochim Biophys Acta (BBA)-Mol Basis of Dis 1822(5):685–689

Du AT, Schuff N, Chao LL, Kornak J, Jagust WJ, Kramer JH, Reed BR, Miller BL, Norman D, Chui HC, Weiner MW (2006) Age effects on atrophy rates of entorhinal cortex and hippocampus. Neurobiol Aging 27(5):733–740

Ezekiel F, Chao L, Kornak J, Du AT, Cardenas V, Truran D, Jagust W, Chui H, Miller B, Yaffe K, Schuff N, Weiner M (2004) Comparisons between global and focal brain atrophy rates in normal aging and Alzheimer disease: boundary shift integral versus tracing of the entorhinal cortex and hippocampus. Alzheimer Dis Assoc Disorders 18(4):196

Fahnestock M, Marchese M, Head E, Pop V, Michalski B, Milgram WN, Cotman CW (2010) BDNF increases with behavioral enrichment and an antioxidant diet in the aged dog. Neurobiol Aging 2010

Frost JL, Le KX, Cynis H, Ekpo E, Kleinschmidt M, Palmour RM, Ervin FR, Snigdha S, Cotman CW, Saido TC, Vassar RJ, St George-Hyslop P, Ikezu T, Schilling S, Demuth HU, Lemere CA (2013) Pyroglutamate-3 amyloid-β deposition in the brains of humans, non-human primates, canines, and Alzheimer disease-like transgenic mouse models. Am J Pathol 183(2):369–381

Giaccone G, Verga L, Finazzi M, Pollo B, Tagliavini F, Frangione B, Bugiani O (1990) Cerebral preamyloid deposits and congophilic angiopathy in aged dogs. Neurosci Lett 114(2):178–183

Gold BT, Johnson NF, Powell DK, Smith CD (2012) White matter integrity and vulnerability to Alzheimer's disease: preliminary findings and future directions. Biochim Biophys Acta (BBA)-Mol Basis Dis 1822(3):416–422

Gonzalez-Soriano J, Garcia PM, Contreras-Rodriguez J, Martinez-Sainz P, Rodriguez-Veiga E (2001) Age-related changes in the ventricular system of the dog brain. Ann Anatomy-Anatomischer Anzeiger 183(3):283–291

Haass C, Selkoe DJ (2007) Soluble protein oligomers in neurodegeneration: lessons from the Alzheimer's amyloid β-peptide. Nat Rev Mol Cell Biol 8(2):101–112

Head E (2000) Brain aging in dogs: parallels with human brain aging and Alzheimer's disease. Veterinary Therapeut: Research Applied Veterinary Med 2(3):247–260

Head E (2011) Neurobiology of the aging dog. Age 33(3):485–496

Head E (2013) A canine model of human aging and Alzheimer's disease. Biochim Biophys Acta (BBA)-Mol Basis Dis 1832(9):1384–1389

Head E, Callahan H, Muggenburg BA, Cotman CW, Milgram NW (1998) Visual-discrimination learning ability and β-amyloid accumulation in the dog. Neurobiol Aging 19(5):415–425

Head E, McCleary R, Hahn FF, Milgram NW, Cotman CW (2000) Region-specific age at onset of β-amyloid in dogs. Neurobiol Aging 21(1):89–96

Head E, Liu J, Hagen TM, Muggenburg BA, Milgram NW, Ames BN, Cotman CW (2002) Oxidative damage increases with age in a canine model of human brain aging. J Neurochem 82(2):375–381

Head E, Nukala VN, Fenoglio KA, Muggenburg BA, Cotman CW, Sullivan PG (2009) Effects of age, dietary, and behavioral enrichment on brain mitochondria in a canine model of human aging. Exp Neurol 220(1):171–176

Head E, Pop V, Sarsoza F, Kayed R, Beckett TL, Studzinski CM, Tomic JL, Glabe CG, Murphy MP (2010) Amyloid-β peptide and oligomers in the brain and cerebrospinal fluid of aged canines. J Alzheimer's Dis 20(2):637–646

Heneka MT, Carson MJ, El Khoury J, Landreth GE, Brosseron F, Feinstein DL, Jacobs AH, Wyss-Coray T, Vitorica J, Ransohoff RM, Herrup K, Frautschy SA, Finsen B, Brown GC, Verkhratsky A, Yamanaka K, Koistinaho J, Latz E, Halle A, Petzold GC, Town T, Morgan D, Shinohara ML, Perry VH, Holmes C, Bazan NG, Brooks DJ, Hunot S, Joseph B, Deigendesch N, Garaschuk O, Boddeke E, Dinarello CA, Breitner JC, Cole GM, Golenbock DT, Kummer MP (2015) Neuroinflammation in Alzheimer's disease. Lancet Neurol 14(4):388–405

Herzig MC, Nostrand WE, Jucker M (2006) Mechanism of cerebral β-amyloid angiopathy: murine and cellular models. Brain Pathol 16(1):40–54

Hwang IK, Yoo KY, Li H, Choi JH, Kwon YG, Ahn Y, Lee IS, Won MH (2007) Differences in doublecortin immunoreactivity and protein levels in the hippocampal dentate gyrus between adult and aged dogs. Neurochem Res 32(9):1604–1609

Hwang IK, Yoon YS, Yoo KY, Li H, Choi JH, Kim DW, Yi SS, Seong JK, Lee IS, Won MH (2008) Differences in lipid peroxidation and Cu, Zn-superoxide dismutase in the hippocampal CA1 region between adult and aged dogs. J Veterinary Med Sci 70(3):273–277

Intlekofer KA, Cotman CW (2013) Exercise counteracts declining hippocampal function in aging and Alzheimer's disease. Neurobiol Dis 57:47–55

Ishihara T, Gondo T, Takahashi M, Uchino F, Ikeda SI, Allsop D, Imai K (1991) Immunohistochemical and immunoelectron microscopical characterization of cerebrovascular and senile plaque amyloid in aged dogs' brains. Brain Res 548(1):196–205

Johnstone EM, Chaney MO, Norris FH, Pascual R, Little SP (1991) Conservation of the sequence of the Alzheimer's disease amyloid peptide in dog, polar bear and five other mammals by cross-species polymerase chain reaction analysis. Mol Brain Res 10(4):299–305

Kayed R, Head E, Thompson JL, McIntire TM, Milton SC, Cotman CW, Glabe CG (2003) Common structure of soluble amyloid oligomers implies common mechanism of pathogenesis. Science 300(5618):486–489

Kiatipattanasakul W, Nakamura SI, Kuroki K, Nakayama H (1997) Immunohistochemical detection of anti-oxidative stress enzymes in the dog brain. Neuropathology 17(4):307–312

Kimotsuki T, Nagaoka T, Yasuda M, Tamahara S, Matsuki N, Ono K (2005) Changes of magnetic resonance imaging on the brain in beagle dogs with aging. J Veterinary Med Sci 67(10):961–967

Lesné S, Koh MT, Kotilinek L, Kayed R, Glabe CG, Yang A, Ashe KH (2006) A specific amyloid-β protein assembly in the brain impairs memory. Nature 440(7082):352–357

Opii WO, Joshi G, Head E, Milgram NW, Muggenburg BA, Klein JB, Pierce WM, Cotman CW, Butterfield DA (2008) Proteomic identification of brain proteins in the canine model of human aging following a long-term treatment with antioxidants and a program of behavioral enrichment: relevance to Alzheimer's disease. Neurobiol Aging 29(1):51–70

Ozawa M, Chambers JK, Uchida K, Nakayama H (2016) The relation between canine cognitive dysfunction and age-related brain lesions. J Veterinary Med Sci 78(6):997–1006

Papaioannou N, Tooten PC, van Ederen AM, Bohl JR, Rofina J, Tsangaris T, Gruys E (2001) Immunohistochemical investigation of the brain of aged dogs. I. Detection of neurofibrillary tangles and of 4-hydroxynonenal protein, an oxidative damage product, in senile plaques. Amyloid 8(1):11–21

Pekcec A, Baumgärtner W, Bankstahl JP, Stein VM, Potschka H (2008) Effect of aging on neurogenesis in the canine brain. Aging Cell 7(3):368–374

Prior R, D'Urso D, Frank R, Prikulis I, Pavlakovic G (1996) Loss of vessel wall viability in cerebral amyloid angiopathy. Neuroreport 7(2):562–564

Pugliese M, Geloso MC, Carrasco JL, Mascort J, Michetti F, Mahy N (2006) Canine cognitive deficit correlates with diffuse plaque maturation and S100β (−) astrocytosis but not with insulin cerebrospinal fluid level. Acta Neuropathol 111(6):519–528

Rofina JE, Singh K, Skoumalova-Vesela A, van Ederen AM, van Asten AJ, Wilhelm J, Gruys E (2004) Histochemical accumulation of oxidative damage products is associated with Alzheimer-like pathology in the canine. Amyloid 11(2):90–100

Rofina JE, Van Ederen AM, Toussaint MJM, Secreve M, Van Der Spek A, Van Der Meer I, Eerdenburg FJ, Gruys E (2006) Cognitive disturbances in old dogs suffering from the canine counterpart of Alzheimer's disease. Brain Res 1069(1):216–226

Romanucci M, Della Salda L (2015) Oxidative stress and protein quality control systems in the aged canine brain as a model for human neurodegenerative disorders. Oxidative Med Cell Longevity 2015:e940131

Sarasa L, Allué JA, Pesini P, González-Martínez Á, Sarasa M (2013) Identification of β-amyloid species in canine cerebrospinal fluid by mass spectrometry. Neurobiol Aging 34(9): 2125–2132

Satou T, Cummings BJ, Head E, Nielson KA, Hahn FF, Milgram NW, Velazquez P, Cribbs DH, Tenner AJ, Cotman CW (1997) The progression of β-amyloid deposition in the frontal cortex of the aged canine. Brain Res 774(1):35–43

Schutt T, Helboe L, Pedersen LO, Waldemar G, Berendt M, Pedersen JT (2016) Dogs with cognitive dysfunction as a spontaneous model for early Alzheimer's disease: a translational study of neuropathological and inflammatory markers. J Alzheimer's Dis 52(2):433–449

Selkoe DJ (1994) Normal and abnormal biology of the beta-amyloid precursor protein. Ann Rev Neurosci 17(1):489–517

Selkoe DJ (2008) Soluble oligomers of the amyloid β-protein impair synaptic plasticity and behavior. Behav Brain Res 192(1):106–113

Selkoe DJ, Bell DS, Podlisny MB, Price DL, Cork LC (1987) Conservation of brain amyloid proteins in aged mammals and humans with Alzheimer's disease. Science 235(4791):873–877

Shigenaga MK, Hagen TM, Ames BN (1994) Oxidative damage and mitochondrial decay in aging. Proc Natl Acad Sci 91(23):10771–10778

Shimada A, Kuwamura M, Akawkura T, Umemura T, Takada K, Ohama E, Itakura C (1992) Topographic relationship between senile plaques and cerebrovascular amyloidosis in the brain of aged dogs. J Veterinary Med Sci 54(1):137–144

Šimić G, Kostović I, Winblad B, Bogdanović N (1997) Volume and number of neurons of the human hippocampal formation in normal aging and Alzheimer's disease. J Comparative Neurol 379(4):482–494

Siwak-Tapp CT, Head E, Muggenburg BA, Milgram NW, Cotman CW (2007) Neurogenesis decreases with age in the canine hippocampus and correlates with cognitive function. Neurobiol Learning Memory 88(2):249–259

Siwak-Tapp CT, Head E, Muggenburg BA, Milgram NW, Cotman CW (2008) Region specific neuron loss in the aged canine hippocampus is reduced by enrichment. Neurobiol Aging 29(1):39–50

Skoumalova A, Rofina J, Schwippelova Z, Gruys E, Wilhelm J (2003) The role of free radicals in canine counterpart of senile dementia of the Alzheimer type. Exp Gerontol 38(6):711–719

Snigdha S, de Rivera C, Milgram NW, Cotman C (2014) Exercise enhances memory consolidation in the aging brain. Front Aging Neurosci 6:3

Su MY, Head E, Brooks WM, Wang Z, Muggenberg BA, Adam GE, Sutherland RJ, Cotman CW, Nalcioglu O (1998) Magnetic resonance imaging of anatomic and vascular characteristics in a canine model of human aging. Neurobiol Aging 19(5):479–485

Swanson KS, Vester BM, Apanavicius CJ, Kirby NA, Schook LB (2007) Implications of age and diet on canine cerebral cortex transcription. Neurobiol Aging 30(8):1314–1326

Tapp PD, Siwak CT, Gao FQ, Chiou JY, Black SE, Head E, Muggenburg BA, Cotman CW, Milgram NW, Su MY (2004) Frontal lobe volume, function, and β-amyloid pathology in a canine model of aging. J Neurosci 24(38):8205–8213

Thal DR, Rüb U, Orantes M, Braak H (2002) Phases of Aβ-deposition in the human brain and its relevance for the development of AD. Neurology 58(12):1791–1800

Uchida K, Miyauchi Y, Nakayama H, Goto N (1990) Amyloid angiopathy with cerebral hemorrhage and senile plaque in aged dogs. Nippon Juigaku Zasshi 52(3):605–611

Uchida K, Nakayama H, Goto N (1991) Pathological studies on cerebral amyloid angiopathy, senile plaques and amyloid deposition in visceral organs in aged dogs. J Veterinary Med Sci 53(6):1037–1042

Uchida K, Tani Y, Uetsuka K, Nakayama H, Goto N (1992) Immunohistochemical studies on canine cerebral amyloid angiopathy and senile plaques. J Veterinary Med Sci 54(4):659–667

Uchida K, Okuda R, Yamaguchi R, Tateyama S, Nakayama H, Goto N (1993) Double-labeling immunohistochemical studies on canine senile plaques and cerebral amyloid angiopathy. J Veterinary Med Sci 55(4):637–642

Uchida K, Kuroki K, Yoshino T, Yamaguchi R, Tateyama S (1997) Immunohistochemical study of constituents other than β-protein in canine senile plaques and cerebral amyloid angiopathy. Acta Neuropathol 93(3):277–284

Walsh DM, Klyubin I, Fadeeva JV, Rowan MJ, Selkoe DJ (2002) Amyloid-β oligomers: their production, toxicity and therapeutic inhibition. Biochem Soc Trans 30(4):552–557

West MJ (1993) Regionally specific loss of neurons in the aging human hippocampus. Neurobiol Aging 14(4):287–293

West MJ, Kawas CH, Martin LJ, Troncoso JC (2000) The CA1 region of the human hippocampus is a hot spot in Alzheimer's disease. Ann NY Acad Sci 908(1):255–259

Wilcock DM (2013) Neuroinflammatory phenotypes and their roles in Alzheimer's disease. Neurodegenerative Dis 13(2-3):183–185

Wisniewski H, Johnson AB, Raine CS, Kay WJ, Terry RD (1970) Senile plaques and cerebral amyloidosis in aged dogs. A histochemical and ultrastructural study. Lab Investig 23:287–296

Wisniewski HM, Wegiel J, Morys J, Bancher C, Soltysiak Z, Kim KS (1990) Aged dogs: an animal model to study beta-protein amyloidogenesis. In: Alzheimer's disease. Epidemiology, neuropathology, neurochemistry, and clinics. Springer, New York, pp 151–167

Yoshino T, Uchida K, Tateyama S, Yamaguchi R, Nakayama H, Goto N (1996) A retrospective study of canine senile plaques and cerebral amyloid angiopathy. Veterinary Pathol Online 33(2):230–234

Neuropathology of Feline Dementia

6

Elizabeth Head and Danièlle Gunn-Moore

Aging cats show behavioral changes that may indicate neurodegenerative processes. In this chapter, we provide an overview of brain changes that have been reported as cats age. The primary focus has been on pathologies that are also seen in human and canine brain, including beta-amyloid plaques and neurofibrillary tangles. There are several similarities between the neuropathology of aging cat brain and that of canine and human brain. Interestingly, there are also unique features of feline brain aging that suggest they may be a useful animal model for understanding Alzheimer's disease. There are many gaps in our knowledge related to which neurodegenerative changes are associated with cognitive decline in aging cats, suggesting the need for further studies in characterized animals.

6.1 Introduction

Nutritional and veterinarian care improvements have led to greater life expectancy in cats, such that the percentage of pet cats of over 7 years of age in the USA has increased to over 40% and more than 10% are over 12 years of age (Laflamme and Gunn-Moore 2014). Aging cats are living to much older ages now, reaching a median longevity of 14 years (O'Neill et al. 2015). As with most aging mammals, increasing life expectancy is associated with neurodegeneration. As described in previous chapters, aging cats can develop cognitive impairment. The purpose of this chapter is to

E. Head (✉)
Department of Pharmacology and Nutritional Sciences, Sanders-Brown Center on Aging, University of Kentucky, Lexington, KY 40536, USA
e-mail: Elizabeth.Head@uky.edu

D. Gunn-Moore
The Royal (Dick) School of Veterinary Studies and The Roslin Institute, The University of Edinburgh, Easter Bush Campus, Roslin EH25 9RG, Scotland
e-mail: Danielle.Gunn-Moore@ed.ac.uk

G. Landsberg et al. (eds.), *Canine and Feline Dementia*,
DOI 10.1007/978-3-319-53219-6_6

provide an overview of brain changes that can occur in cats as they age, and how these changes may underlie features observed by pet owners and clinicians.

There are few studies of the aged cat brain. However, those that have been performed have shown that aging cats can show some features of brain aging that are similar to dogs and to humans but with some important differences (Gunn-Moore et al. 2007; Vite and Head 2014).

6.2 Structural Brain Changes and Neuron Losses

Human and canine brain aging are associated with gradual brain atrophy, region-specific neuron loss, and reduced synapse-associated proteins (Head 2011; Juraska and Lowry 2012; West 1993). In Alzheimer's disease (AD), these losses can be extensive and are associated with the development of dementia (Scheff et al. 2006; West et al. 1994).

Aged cats also show morphological brain changes that may signal neuron loss and brain atrophy (Fig. 6.1). Neuron loss in the molecular layer of the cerebellum is observed in 12 – 13-year-old cats as compared to 2 – 3-year-old animals (Zhang et al. 2006). Dendritic loss may also occur in the Purkinje cells of the cerebellum, as there is a decrease in the amount of immunolabeling for neurofilaments in older cats (Zhang et al. 2006).

Similarly, there may be a loss of dendrites in the caudate nucleus of aged cats (over 10 years of age) based upon ultrastructural studies and synaptic impairments on the basis of electrophysiological experiments (Levine 1988; Levine et al. 1986, 1988). Dysfunction in the caudate nucleus may result in impairments to motor function and/or in the ability to habituate to repeated stimuli (Levine et al. 1987; Villablanca et al. 1978). Certainly, many older cats pace obsessively, and the author (DGM) has identified a number of cats with cognitive dysfunction syndrome and no other brain pathology that demonstrated marked parallel pacing in the latter part of their life (DGM, personal observation). The inability to habituate to repeated stimuli

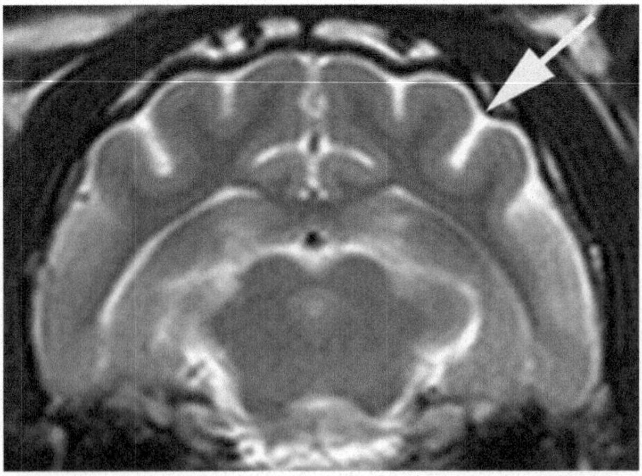

Fig. 6.1 MRI of an aged cat brain showing widening of the sulci (*arrow*) indicating brain atrophy

may explain why cats with cognitive dysfunction syndrome are easily stressed by stimuli, even when they were previously accustomed to it, e.g., the noise of a vacuum cleaner or the doorbell ringing (DGM, personal observation).

Neuron loss is also observed in the hippocampus of aging cats, particularly after the age of 14 years, when both Aβ and tau phosphorylation are present (Chambers et al. 2015). This may be particularly important as the hippocampus is essential for learning and memory.

6.3 Aβ Accumulation

The beta-amyloid protein (Aβ) is derived from a longer precursor (amyloid precursor protein) and has been linked to the development of AD (Hardy and Selkoe 2002). Aβ is deposited in extracellular plaques in the brain, with the 40 – 42 amino acid long peptide predominating. Aβ plaques are one of the two defining features in the human brain that must be present in sufficient numbers for a final diagnosis of AD (Hardy and Selkoe 2002; Mirra 1997; Selkoe 1996).

Aβ neuropathology in aging cats shows some similarities to humans and other animals (e.g., dog) but some interesting differences as well. The sequence of Aβ in cats has one amino acid difference from humans (Brinkmalm et al. 2012), and this may lead to differences in the type of Aβ that accumulates with age. Although Aβ in plaques has been reported in several studies, increasing with prevalence in cats over 8 years of age (Brellou et al. 2005; Cummings et al. 1996; Gunn-Moore et al. 2006; Head et al. 2005; Kuroki et al. 1997; Nakamura et al. 1996; Takeuchi et al. 2008), this has been most commonly reported using antibodies against the C-terminus of Aβ (Aβ42 specific) or using antibodies that recognize Aβ17-24. In cats, amyloid plaques appear to be even more diffuse than those seen in dogs and quite unlike the well-developed and circumscribed amyloid plaques that are typical of AD in humans (Chambers et al. 2015; Cummings et al. 1996; Gunn-Moore et al. 2006; Head et al. 2005) (Fig. 6.2).

Interestingly, very little Aβ1-16 or Aβ1-40 is observed (Head et al. 2005), and truncated Aβ (AβpN3) found in dog and human brain has not been detected in aged cat brain (Chambers et al. 2011). The lack of more soluble Aβ1-40 suggests that after deposition within the extracellular space, it is rapidly turned over (Hyman et al. 1993), suggesting that plaque accumulation in elderly cat brains is more similar to that seen in non-demented elderly humans, than it is to people with AD (Head et al. 2005; Iwatsubo et al. 1994). Posttranslationally modified Aβ (isomerized Aβ), which may represent "older" deposits of Aβ in plaques, is observed in human brain (Fonseca et al. 1999) and dog brain (Satou et al. 1997) but not cat brain (Head et al. 2005).

A soluble more toxic conformation of Aβ, called oligomers, is thought to be a key culprit causing synaptic dysfunction in the AD brain (Glabe 2006). These small conformations of Aβ can diffuse from the synapse and aggregate further into protofibrils (Walsh and Selkoe 2004). Recently, Aβ oligomers have been described in cats where they are observed in animals over 8 years of age (Chambers et al. 2015).

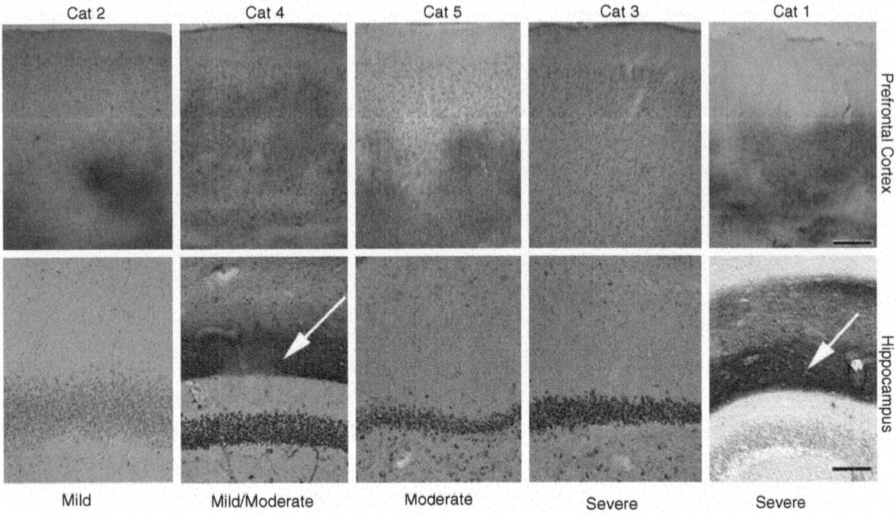

Fig. 6.2 Representative immunostaining for Aβ17-24 (4G8) in all five clinically characterized aged cats from the prefrontal cortex (bar = 100 μm) and hippocampus (bar = 75 μm) shown in order of progressively more severe signs of cognitive decline. Note the individual variability in the presence or absence of outer molecular layer Aβ accumulation as well as diffuse amyloid deposition. No consistent association between the severity of clinical signs of cognitive dysfunction and the extent of Aβ deposition was observed (Picture reprinted with permission Head et al. 2005)

In cats, the link between Aβ oligomers and their possible role in synaptic dysfunction or behavioral changes has yet to be established. Two studies that looked at a small number of cats ($n = 20$) with abnormal behavioral signs found diffuse amyloid plaques in older individuals (greater than 10 years of age) (Cummings et al. 1996; Gunn-Moore et al. 2006; Head et al. 2005). However, these findings were not consistent, and a study of five cats with well-defined signs of cognitive dysfunction found that the extent of the Aβ deposition did not correlate well with the severity of the behavioral changes (Head et al. 2005).

6.4 Tau Pathology

The intracellular accumulation of abnormally hyperphosphorylated tau can result in development of neurofibrillary tangles, which are one of the two classical features of AD (Trojanowski et al. 1993), and correlate with the severity of dementia (Arriagada et al. 1992; Nelson et al. 2007).

It is fascinating to note that there are several descriptions of tau abnormalities in the aged cat brain that are similar to the hyperphosphorylated tau protein that accumulates as neurofibrillary tangles in the AD brain (Chambers et al. 2015; Gunn-Moore et al. 2006; Head et al. 2005) (Fig. 6.3). Chambers et al. (2015) show

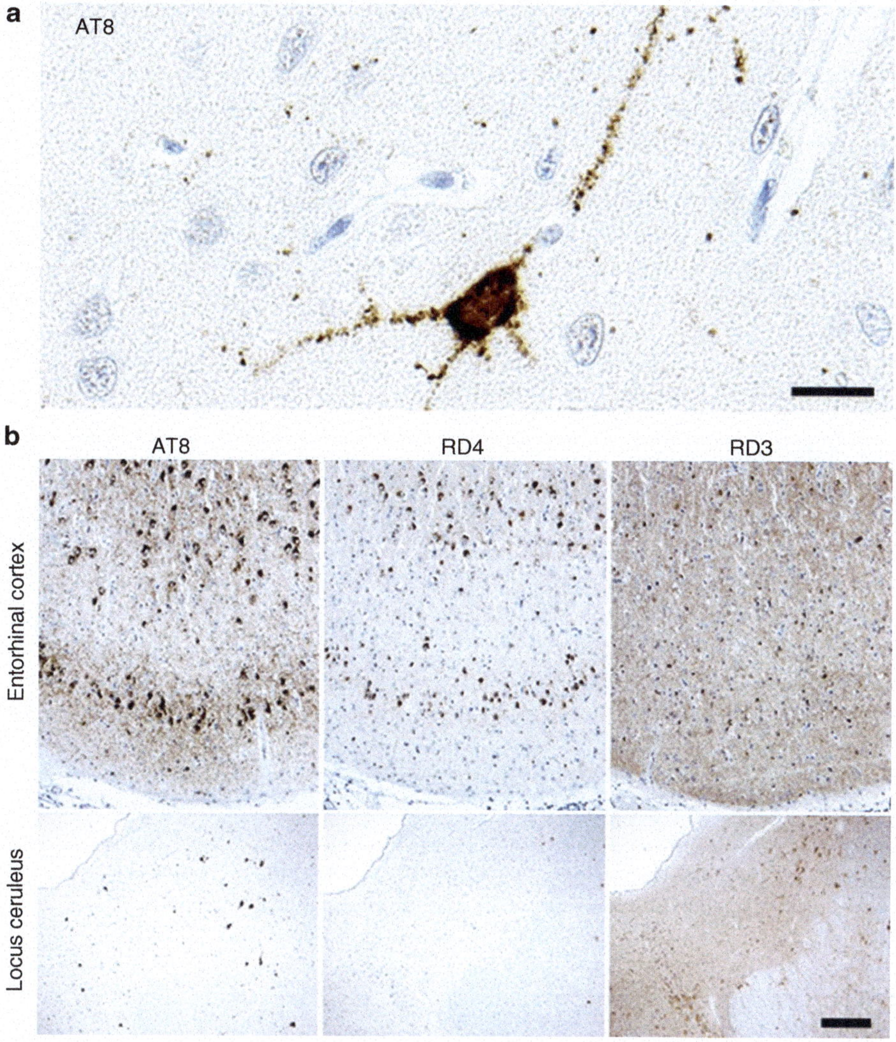

Fig. 6.3 Hyperphosphorylated tau accumulation in the entorhinal cortex and locus coeruleus of cat brains. (**a**) Immunohistochemistry of the entorhinal cortex of a cat with mild hyperphosphorylated tau accumulation (15-year-old, case no. 13) (AT8). The neuronal soma and dendrites are positively stained for hyperphosphorylated tau. Bar = 20 μm. (**b**) Immunohistochemistry of the entorhinal cortex and locus coeruleus of a cat with severe hyperphosphorylated tau accumulation (14-year-old, case no. 12) for AT8, 3-repeat tau (RD3), and 4-repeat tau (RD4). AT8-positive aggregates are also positively stained for 3-repeat tau and 4-repeat tau on consecutive sections. Bar = 300 μm (Reprinted with permission from Chambers et al. 2015)

nice evidence of silver staining (Gallyas-Braak) and neurons containing hyperphosphorylated tau Ser202/Thr205 (AT8) in the hippocampus of aging cats (of greater than 14 years of age) (Chambers et al. 2015). However, the morphology of these intracellular tau accumulations is similar, albeit not identical, to tau lesions detected in the human AD brain, but may represent early forms of tangle pathology.

To date, while hyperphosphorylated tau has been seen in elderly cats with behavioral and/or neurological disease (Chambers et al. 2015; Gunn-Moore et al. 2006), its role in cognitive dysfunction syndrome cannot be determined as too few cats have yet been studied.

6.5 Neurotransmitter Systems

The cholinergic hypothesis for AD suggests, on the basis of autopsy studies, that a loss of cholinergic neurons accounts for decreased acetylcholine at the synapse and thus impaired cognition (Bartus et al. 1982; Coyle et al. 1983). This hypothesis led to the development of four of the five currently approved drugs used to treat patients with AD, all of which provide benefits by inhibiting cholinesterase and consequently increasing acetylcholine levels (Neugroschl and Sano 2010).

The noradrenergic locus coeruleus of older cats also shows losses in the size of choline acetyltransferase (ChAT)-bearing neurons, with fewer microfilaments within their dendrites, and some cats showing axonal degeneration and myelin disruption, suggesting a similar impairment of the cholinergic system occurs with age in cats (Zhang et al. 2005). These abnormalities in the cholinergic system may result in disruption in the sleep-wake cycle (Chase 1983), something that is seen in many cats with cognitive dysfunction syndrome. There have been few studies on changes in other neurotransmitter systems (e.g., dopamine, serotonin, etc.) in the brains of aging cats, and this represents an area requiring further study.

6.6 Vascular Pathology

Vascular pathology is rapidly being recognized as a key contributor to cognitive decline in humans and is often comorbid with AD neuropathology (Snyder et al. 2015). Specifically, cerebral amyloid angiopathy is thought to mediate vascular dysfunction in human and canine brain (Attems 2005; Attems et al. 2005; Ellis et al. 1996; Head 2011). The observation of cerebral amyloid angiopathy (CAA), a form of vascular pathology, has been somewhat variable in aging cat brain studies. Several studies have observed CAA (Fig. 6.4) (Cummings et al. 1996; Gunn-Moore et al. 2006; Head et al. 2005; Nakamura et al. 1996). However, in a more recent study of 25 cats ranging in age from 50 days to 22 years of age, no vascular Aβ was observed (Chambers et al. 2015). The lack of concordance in these two studies suggests that additional studies with more animals may be interesting.

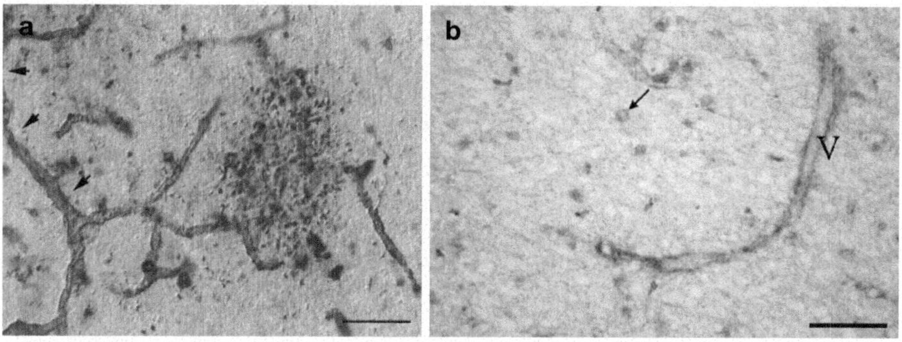

Fig. 6.4 Aged cat brains also show evidence of Aβ in association with small blood vessels (*arrows*), typically in the cortical white matter (**a**). (**b**) shows evidence of vascular pathology (V) in white matter along with scattered neuronal labeling (*arrow*). Bars = 50 μm. (**a**) is reproduced with permission from Head et al. 2005

6.7 Summary

Aged cats develop neuropathology that in many respects overlaps with that observed in aging dogs and in aging people. Uniquely for cats, there appears to be a significant tau phosphorylation with age that may represent early neurofibrillary pathology. Other aspects of brain aging (e.g., oxidative damage) are, as yet, still unexplored in these animals. Given the occurrence of cognitive decline or behavioral changes with aging in cats, they represent an animal model that may provide some unique features that can facilitate the development of novel interventions for AD.

Acknowledgments The images, reprinted with permission from the *Journal of Feline Medicine and Surgery*, were generated by a study that was funded by Petsavers.

References

Arriagada PV, Growdon JH, Hedley-Whyte ET, Hyman BT (1992) Neurofibrillary tangles but not senile plaques parallel duration and severity of Alzheimer's disease. Neurology 42(3):631–631

Attems J (2005) Sporadic cerebral amyloid angiopathy: pathology, clinical implications, and possible pathomechanisms. Acta Neuropathol 110(4):345–359

Attems J, Jellinger KA, Lintner F (2005) Alzheimer's disease pathology influences severity and topographical distribution of cerebral amyloid angiopathy. Acta Neuropathol 110(3):222–231

Bartus RT, Dean R III, Beer B, Lippa AS (1982) The cholinergic hypothesis of geriatric memory dysfunction. Science 217(4558):408–414

Brellou G, Vlemmas I, Lekkas S, Papaioannou N (2005) Immunohistochemical investigation of amyloid ß-protein (Aß) in the brain of aged cats. J Histol Histopathol 20(3):725–731

Brinkmalm G, Portelius E, Öhrfelt A, Mattsson N, Persson R, Gustavsson MK et al (2012) An online nano-LC-ESI-FTICR-MS method for comprehensive characterization of endogenous fragments from amyloid β and amyloid precursor protein in human and cat cerebrospinal fluid. J Mass Spectrom 47(5):591–603

Chambers JK, Mutsuga M, Uchida K, Nakayama H (2011) Characterization of AβpN3 deposition in the brains of dogs of various ages and other animal species. Amyloid 18(2):63–71

Chambers JK, Tokuda T, Uchida K, Ishii R, Tatebe H, Takahashi E et al (2015) The domestic cat as a natural animal model of Alzheimer's disease. Acta Neuropathol Commun 3(1):1

Chase MH (1983) Synaptic mechanisms and circuitry involved in motoneuron control during sleep. Int Review Neurobiol 24:213–258

Coyle JT, Price DL, DeLong MR (1983) Alzheimer's disease: a disorder of cortical cholinergic innervation. Science 219(4589):1184–1190

Cummings BJ, Satou T, Head E, Milgram NW, Cole GM, Savage MJ et al (1996) Diffuse plaques contain C-terminal Aβ 42 and not Aβ 40: evidence from cats and dogs. Neurobiol Aging 17(4):653–659

Ellis RJ, Olichney JM, Thal LJ, Mirra SS, Morris JC, Beekly D, Heyman A (1996) Cerebral amyloid angiopathy in the brains of patients with Alzheimer's disease The CERAD experience, part XV. Neurology 46(6):1592–1596

Fonseca MI, Head E, Velazquez P, Cotman CW, Tenner AJ (1999) The presence of isoaspartic acid in β-amyloid plaques indicates plaque age. Exp Neurol 157(2):277–288

Glabe CG (2006) Common mechanisms of amyloid oligomer pathogenesis in degenerative disease. Neurobiol Aging 27(4):570–575

Gunn-Moore DA, McVee J, Bradshaw JM, Pearson GR, Head E, Gunn-Moore FJ (2006) Ageing changes in cat brains demonstrated by β-amyloid and AT8-immunoreactive phosphorylated tau deposits. J Feline Med Surg 8(4):234–242

Gunn-Moore D, Moffat K, Christie LA, Head E (2007) Cognitive dysfunction and the neurobiology of ageing in cats. J Small Animal Pract 48(10):546–553

Hardy J, Selkoe DJ (2002) The amyloid hypothesis of Alzheimer's disease: progress and problems on the road to therapeutics. Science 297(5580):353–356

Head E (2011) Neurobiology of the aging dog. Age 33(3):485–496

Head E, Moffat K, Das P, Sarsoza F, Poon WW, Landsberg G, Cotman CW, Murphy MP (2005) β-Amyloid deposition and tau phosphorylation in clinically characterized aged cats. Neurobiol Aging 26(5):749–763

Hyman BT, Marzloff K, Arriagada PV (1993) The lack of accumulation of senile plaques or amyloid burden in Alzheimer's disease suggests a dynamic balance between amyloid deposition and resolution. J Neuropathol Exp Neurol 52(6):594–600

Iwatsubo T, Odaka A, Suzuki N, Mizusawa H, Nukina N, Ihara Y (1994) Visualization of Aβ42 (43) and Aβ40 in senile plaques with end-specific Aβ monoclonals: evidence that an initially deposited species is Aβ42 (43). Neuron 13(1):45–53

Juraska JM, Lowry NC (2012) Neuroanatomical changes associated with cognitive aging. Curr Topics Behav Neurosci 10:137–162

Kuroki K, Uchida K, Kiatipattanasakul W, Nakamura SI, Yamaguchi R, Nakayama H, Tateyama S (1997) Immunohistochemical detection of tau protein in various non-human animal brains. Neuropathology 17(3):174–180

Laflamme D, Gunn-Moore D (2014) Nutrition of aging cats. Veterinary Clin N Am: Small Animal Pract 44(4):761–774

Levine MS (1988) Neurophysiological and morphological alterations in caudate neurons in aged cats. Ann N Y Acad Sci 515(1):314–328

Levine MS, Adinolfi AM, Fisher RS, Hull CD, Buchwald NA, McAllister JP (1986) Quantitative morphology of medium-sized caudate spiny neurons in aged cats. Neurobiol Aging 7(4):277–286

Levine MS, Lloyd RL, Fisher RS, Hull CD, Buchwald NA (1987) Sensory, motor and cognitive alterations in aged cats. Neurobiol Aging 8:253–263

Levine MS, Adinolfi AM, Fisher RS, Hull CD, Guthrie D, Buchwald NA (1988) Ultrastructural alterations in caudate nucleus in aged cats. Brain Res 440(2):267–279

Mirra SS (1997) The CERAD neuropathology protocol and consensus recommendations for the postmortem diagnosis of Alzheimer's disease: a commentary. Neurobiol Aging 18(4):S91–S94

Nakamura SI, Nakayama H, Kiatipattanasakul W, Uetsuka K, Uchida K, Goto N (1996) Senile plaques in very aged cats. Acta Neuropathol 91(4):437–439

Nelson PT, Jicha GA, Schmitt FA, Liu H, Davis DG, Mendiondo MS, Abner EL, Markesbery WR (2007) Clinicopathologic correlations in a large Alzheimer disease center autopsy cohort: neuritic plaques and neurofibrillary tangles "do count" when staging disease severity. J Neuropathol Exp Neurol 66(12):1136–1146

Neugroschl J, Sano M (2010) Current treatment and recent clinical research in Alzheimer's disease. Mount Sinai J Med 77(1):3–16

O'Neill DG, Church DB, McGreevy PD, Thomson PC, Brodbelt DC (2015) Longevity and mortality of cats attending primary care veterinary practices in England. J Feline Med Surg 17(2):125–133

Satou T, Cummings BJ, Head E, Nielson KA, Hahn FF, Milgram NW, Velazquez P, Cribbs DH, Tenner AJ, Cotman CW (1997) The progression of β-amyloid deposition in the frontal cortex of the aged canine. Brain Res 774(1):35–43

Scheff SW, Price DA, Schmitt FA, Mufson EJ (2006) Hippocampal synaptic loss in early Alzheimer's disease and mild cognitive impairment. Neurobiol Aging 27(10):1372–1384

Selkoe DJ (1996) Amyloid β-protein and the genetics of Alzheimer's disease. J Biol Chem 271(31):18295–18298

Snyder HM, Corriveau RA, Craft S, Faber JE, Greenberg SM, Knopman D, Lamb BT, Montine TJ, Nedergaard M, Schaffer CB, Schneider JA, Wellington C, Wilcock DM, Zipfel GJ, Zlokovic B, Bain LJ, Bosetti F, Galis ZS, Koroshetz W, Carrillo MC (2015) Vascular contributions to cognitive impairment and dementia including Alzheimer's disease. Alzheimer's Dementia 11(6): 710–717

Takeuchi Y, Uetsuka K, Murayama M, Kikuta F, Takashima A, Doi K, Nakayama H (2008) Complementary distributions of amyloid-β and neprilysin in the brains of dogs and cats. Veterinary Pathol Online 45(4):455–466

Trojanowski JQ, Schmidt ML, Shin RW, Bramblett GT, Rao D, Lee VMY (1993) Altered tau and neurofilament proteins in neuro-degenerative diseases: diagnostic implications for Alzheimer's disease and lewy body dementias. Brain Pathol 3(1):45–54

Villablanca JR, Olmstead CE, de Andres I (1978) Effects of caudate nuclei or frontal cortical ablations in kittens: responsiveness to auditory stimuli and comparisons with adult-operated littermates. Exp Neurol 61(3):635–649

Vite CH, Head E (2014) Aging in the canine and feline brain. Veterinary Clin N Am: Small Animal Pract 44(6):1113–1129

Walsh DM, Selkoe DJ (2004) Oligomers on the brain: the emerging role of soluble protein aggregates in neurodegeneration. Protein Peptide Lett 11(3):213–228

West MJ (1993) Regionally specific loss of neurons in the aging human hippocampus. Neurobiol Aging 14(4):287–293

West MJ, Coleman PD, Flood DG, Troncoso JC (1994) Differences in the pattern of hippocampal neuronal loss in normal ageing and Alzheimer's disease. Lancet 344(8925):769–772

Zhang JH, Sampogna S, Morales FR, Chase MH (2005) Age-related changes in cholinergic neurons in the laterodorsal and the pedunculo-pontine tegmental nuclei of cats: a combined light and electron microscopic study. Brain Res 1052(1):47–55

Zhang C, Hua T, Zhu Z, Luo X (2006) Age-related changes of structures in cerebellar cortex of cat. J Biosci 31(1):55–60

Proteomic Approaches for Diagnostics of Canine and Feline Dementia

7

Branislav Kovacech, Monika Zilkova, Jozef Hanes, and Rostislav Skrabana

Cognitive dysfunction syndrome (CDS) in pet animals constitutes a pressing problem of the modern society. Millions of senior dogs and cats undergo age-related behavioral changes that impact their social interactions. Currently, it is difficult to discriminate between normal aging and dementing processes. The situation is akin to that in humans; however, human medicine receives enormous resources that resulted in a set of current diagnostic criteria including a number of dementia assessing scales, diagnostic assays, and novel potential biomarkers. While animal well-being is not in the limelight of societal interest, the dementia diagnostics starts to catch up (Madari et al. 2015; Schütt et al. 2015). Nevertheless, biochemical markers related to the animal dementia are underdeveloped, despite the fact that dogs and cats provide natural models for human dementia (Bosch et al. 2012; Chambers et al. 2015).

7.1 Tissue Types for Biomarker Identification and Quantification

Biochemical diagnostics usually focuses on the body fluids as the most accessible sources of biological markers of disease. The concentrations of potential brain-derived biomarkers in the body fluids substantially decrease in the following order: (brain)–cerebrospinal fluid–blood–urine. Saliva and tears complete the list, being positioned somewhere between blood and urine; nonetheless, their diagnostic utilization is not much exploited.

B. Kovacech • M. Zilkova (✉) • J. Hanes • R. Skrabana
Institute of Neuroimmunology, Slovak Academy of Sciences, Dubravska 9, Bratislava, Slovak Republic
e-mail: monika.zilkova@savba.sk; jozef.hanes@gmail.com; rostislav.skrabana@savba.sk

G. Landsberg et al. (eds.), *Canine and Feline Dementia*,
DOI 10.1007/978-3-319-53219-6_7

7.1.1 Cerebrospinal Fluid

Cerebrospinal fluid (CSF) protects the central nervous system (CNS) from physical stress and plays an essential role in homeostasis of the brain and regulation of neuronal functioning. CSF is a clear liquid of extremely low cellularity (0–8 cells/μL). It is secreted by cells of the choroid plexus and ependymal cells that line the ventricles and is absorbed into the venous system in the subarachnoid space (Johanson et al. 2008). Its composition is determined on one hand by the relatively free exchange of proteins, peptides, and metabolites with the brain tissue and on the other by a highly restricted and regulated exchange with blood across the blood-CSF barrier (McComb 1983; Iliff et al. 2012; Tarasoff-Conway et al. 2015).

Since CSF is in direct contact with the brain interstitial fluid (ISF) that soaks the neurons, biochemical changes in the brain are reflected in CSF. Furthermore, CSF has low protease activity, and most proteinaceous molecules do not change upon collection provided, of course, the collected CSF sample is not contaminated by blood. Therefore, CSF should be the best source of biomarkers that reflect the pathological changes of the brain (Blennow et al. 1993).

The total protein concentration reaches around 0.3 mg/mL, which increases with age (Maurer 2010). In dogs and cats, protein concentration >0.8 mg/mL indicates alteration of the blood-brain barrier and/or increased local synthesis. Albumin (~0.2 mg/mL) constitutes around 67% of the total CSF protein; beta-trace, prealbumin, immunoglobulins, and transferrin make up additional 27%. On the other side of the CSF protein, rosters are such neuron-specific proteins like myelin basic protein, S-100, cytokines, and bioactive peptides with concentrations in sub-pg/ml levels. CSF concentrations of dementia-related proteins like amyloid β 1-42 (Aβ42) and tau are around 550 pg/mL (Borghys et al. 2014) and 30 pg/mL (Roerig et al. 2013), respectively.

Thus, the concentrations of proteinaceous molecules in CSF span at least nine orders of magnitude. Eighty percent of CSF proteins, the most abundant ones, are blood derived, and only 20% come from the brain ISF. The current and future dementia markers are expected to be found in the low concentration range of >3000 CSF proteins and peptides, resulting in an enormous difficulties for analytical quantifications (Schutzer et al. 2010; Guldbrandsen et al. 2014).

7.1.2 Blood (Plasma and Serum)

Blood (plasma or serum) is far more accessible than CSF; however, the analytical complexity of plasma (serum) appears even higher than that of CSF. First, the concentration range of the individual proteins spans 10–11 orders of magnitude (Anderson and Anderson 2002). Second, the total plasma protein concentration is ~55–75 mg/mL, 200-fold higher than in CSF, and thus any brain-derived proteins will be highly diluted. For example, Aβ42 is present in the dog plasma at 25–75 pg/mL, tenfold lower than in CSF (Gonzalez-Martinez et al. 2011; Schütt et al. 2015).

Third, proteins and peptides might have a short half-life in plasma due to fast renal clearance. Blood contains relatively high proteolytic and other enzymatic activities causing most intracellular proteins released into the bloodstream to undergo degradation and/or modification by proteases and other enzymes, and, therefore, for most of the biomarker candidates, the half-life in the blood is unknown (Werle and Bernkop-Schnürch 2006).

Nevertheless, major human neurodegenerative diseases have a substantial neuroinflammatory component, and some biomarker signatures seem to be connected to the peripheral immune system (reviewed in Chiam et al. 2015; Zafari et al. 2015). It makes, therefore, sense to exploit plasma (preferred over serum) for identification of novel biomarkers.

7.1.3 Urine, Saliva, and Tears

Some of the brain-derived proteins and peptides might find their way to other body fluids like saliva, tears, or urine. For example, neuronal protein tau associated with neurofibrillary pathology and α-synuclein and DJ-1 proteins connected to Parkinson's disease were detected in saliva samples in humans (Devic et al. 2011; Shi et al. 2011). It remains to be seen, however, whether their amount reflects pathological processes in the brain.

Similarly to the blood, the urine is a fluid relatively easy to collect and thus promising source of biochemical markers. Over 1500 proteins have already been identified in the urine (Adachi et al. 2006; Rodríguez-Suárez et al. 2014), and it is not excluded that some of the more hydrophilic proteins and peptides might get accumulated there due to their fast renal clearance.

7.2 Biomarker Quantification Methods Used in Diagnostics

Biomarkers for human dementia, especially Alzheimer's disease, have been analyzed for almost 20 years now, but none of the tests currently available gives 100% specificity and sensitivity. Nevertheless, some of them have been validated for humans and provide substantial help in a differential diagnosis of neurodegenerative disorders (Dubois et al. 2014; Olsson et al. 2016).

7.2.1 Immunoassays

Immunoassays were historically the first techniques that achieved the sufficient sensitivity and specificity to detect the identified biomarkers in body fluids. The widespread ELISA (enzyme-linked immunosorbent assay) format usually requires two antibodies (sandwich ELISA): one is immobilized on a microtiter plate and serves as a capture antibody to fish out the biomarker from the sample and the other is labeled and serves as a detection antibody for the captured molecule. The signal can be amplified by using a third antibody or using a biochemical amplification system. The ELISA tests for neuronal protein tau and Aβ42 in CSF were first developed in the 1990s (Vandermeeren et al. 1993; Hulstaert et al. 1999). The CSF-based tests INNOTEST® hTAU, INNOTEST® PHOSPHO-TAU$_{(181P)}$, INNOTEST® β-AMYLOID$_{(1–40)}$, and INNOTEST® β-AMYLOID$_{(1–42)}$ (Fujirebio Europe, Ghent, Belgium) remain the most useful and characterized assays in quantifying the biomarkers of human neurodegenerative disorders and can be used in canine and feline diagnosis of dementia. Other immunoassays include those for microglia and astrocyte marker YKL-40, neurofilament light chain, alpha-synuclein, DJ-1 protein, and several others (reviewed in Blennow et al. 2016; Olsson et al. 2016). These tests are used in experimental diagnostics, identification of disease subtypes and staging, differential diagnostics, and stratification of patients in clinical trials (Lleo et al. 2015).

Due to the high homology between human, dog and cat variants of these proteins, many of the assays can be immediately applied to use in animals. However, suitable population studies determining the normal and disease-associated values need to be performed prior to drawing any meaningful diagnostic conclusions.

7.2.2 Multiplex and High-Sensitivity Immunoassays

With the increasing number of ELISA tests, it soon became clear that the sample consumption and labor were exceeding the practical laboratory throughput. Modifications were sought to accommodate more than one analyte per well (e.g., Aβ Triplex assay, U-Plex Chemokine 25-plex, Canine Proinflammatory kit, all from Meso Scale Discovery, Rockville, USA) or to use color-coded (magnetic) beads with immobilized capture antibodies, e.g., xMAP® technology from Luminex, for up to 500 analytes in one assay (Ellington et al. 2010).

Standard ELISA assays and their multiplex modifications have lower limits of quantification around 10–100 pg/mL. This prevents quantification of neuronal proteins in plasma, since e.g. tau protein and S-100 calcium-binding protein B are present in sub-pg/mL levels (Shahim et al. 2014). Therefore, highly sensitive diagnostic assays are being developed.

The measurement of tau and S-100 calcium-binding protein B in blood (plasma) was allowed thanks to an ultrasensitive, digital, bead-based immunoassay called SIMOA (single-molecule array) developed by Quanterix Corp. (Boston, USA) and published recently (Rissin et al. 2010). The assay is actually a standard sandwich ELISA assembled on magnetic beads, which are then transferred to a plate containing ~300,000 microwells for quantification. The wells accommodate only 1 bead, which allows quantification of each individual molecule of the analyte (hence "digital" ELISA) leading to ~1000-fold, increases in sensitivity over the conventional assays. Aβ42 was measured using SIMOA in sera of patients resuscitated after cardiac arrest, and elevated levels were found to correlate with the clinical outcome (Zetterberg et al. 2011). The SIMOA assay was also used for the quantification of tau in serum of these patients, and the elevated levels could have utility for hypoxic brain injury assessment and prediction of cerebral function outcome (Randall et al. 2013). Plasma tau was also assessed with the relation to Alzheimer's disease diagnosis and was found to be mildly elevated but not useful as a diagnostic value (Zetterberg et al. 2013). However, the ultrasensitive measurements of tau in the plasma of people suffering from head injuries (concussions, blasts, etc.) or brain hypoxia proved useful for the assessment of the clinical outcome (Ling et al. 2015; Gren et al. 2016).

Another ultrasensitive platform, single-molecule counting (SMC, from Singulex, Alameda, CA, USA), is a bead-based sandwich immunoassay, where individual fluorescently labeled detection antibody molecules are counted with a confocal detection system. The assay has been used for the quantification of neuronal visinin-like protein-1, neurogranin, and Aβ oligomers in the human CSF and showed correlation with the progression of brain atrophy in Alzheimer's disease (Tarawneh et al. 2012, 2015, 2016; Yang et al. 2015).

In a proximity extension assay (PEA, developed by OLINK Proteomics, Uppsala, Sweden), two antibodies carrying partially overlapping complementary oligonucleotides are used for one biomarker. These antibodies are mixed with the sample; by binding the same molecule they get to close proximity, the oligonucleotides bind each other and allow detection of the signal by quantitative PCR. This setup permits quantification of multiple different analytes in one sample simultaneously (Assarsson et al. 2014). Furthermore, the assay is "homogeneous," which means that no washing steps are involved, the sample is simply mixed with the antibodies, and the signal is measured. As opposed to the heterogeneous ELISA assays, homogeneous assays are much less laborious; they are quicker and less prone to analytical confounders.

Finally, immunomagnetic reduction (IMR) assay has recently been developed by MagQu (MagQu Co. Ltd., Taiwan). This is again a homogeneous assay where magnetic nanoparticles coated with a biomarker-specific antibody are mixed with the sample and exposed to external magnetic fields, which forces the nanoparticles to align. Upon binding to a biomarker molecule, the nanoparticles become larger and consequently are slower to align with the field. The differences in the speed of alignment reflect the biomarker concentration. This novel assay has already been applied to measurements of human Aβ42 and Aβ40 (Chiu et al. 2012) and neuronal protein tau (Chiu et al. 2014) in plasma and showed correlations with the diagnosis of Alzheimer's disease and volumetric and cognitive characteristics of patients, respectively.

Despite the fast progress in immunological methods for biomarker detection, little progress has been made over the past 15 years toward a simple and definite dementia diagnostics. This is caused, in major part, by the lack of suitable, disease-specific biomarkers. Many immunoassays are based on the same set of antibodies, just applied to different technological platforms, improving analytical sensitivity but not specificity and selectivity of the assay. Therefore, the hunt is on for novel antibodies with suitable properties detecting other disease-associated (and disease-specific) epitopes on known proteins.

New biomarkers (disease-associated protein variants) are desperately sought for by applying the state-of-the-art proteomic techniques. It is expected that the biomarkers will arise from the better understanding of the pathophysiological processes underlying the progression of different forms of dementias and might thus provide not only improved diagnosis but also novel targets for disease-modifying therapies.

7.3 Proteomic Approaches in Biomarker Discovery

Identification of proteomic biomarkers related to any disorder means comparing the protein complements (proteomes) of normal and diseased tissues and selecting the proteins that are different. Those are candidate biomarkers, which need to be validated in disease models and population studies. It may sound like a straightforward procedure; however, this is exactly what hundreds of scientists were trying to do for cancer over the last more than 30 years with mixed success. The problems encountered are multiple. First, the biomarker for a given disease is usually present at a very low concentration, especially when the disease is in its initial stage and when a possible treatment is most effective. Second, identification of the biomarkers is dependent on the well-defined cohorts of patients and healthy individuals. Selection of such cohorts is often a difficult task, since the diagnostic methods for neurodegenerative diseases are not fully reliable (especially not in the early disease stages) (Beach et al. 2012). Third, the proteomic methods are themselves error prone, often introducing unintentional biases or false-positive results. Fourth, the proteins are present at highly differing amounts, and concentration ranges span up to 12 orders of magnitude. Fifth, the sheer number of various forms of proteins is staggering and overwhelms the capacity of any currently known separation and identification technique. The problem can be explained on neuronal protein tau, whose disease forms generate neurofibrillary pathology. The protein is coded by one gene, whose mRNA is alternatively spliced to produce at least six different protein isoforms. These isoforms are further modified by posttranslational modifications. More than 50 phosphorylation sites have been identified on the protein, along with several glycation, glycosylation, and acetylation sites. These modifications are present in various combinations. On top of these physiological modifications, tau undergoes hyperphosphorylation and multiple abnormal truncations in neurofibrillary pathology (reviewed in Kovacech and Novak 2010, Zilka et al. 2012). Tau, therefore, exists in tens or hundreds of different forms in normal brain, and this number is further multiplied in a diseased tissue. The same protein species scheme applies to all proteins, only differing in the type and number of modifications. These different protein varieties are expected to have (at least slightly) different biological activities. They were termed *proteoforms* by Kelleher's group, which denote highly related protein molecules arising from all combinatorial sources of variation giving rise to products arising from a single gene (Smith and Kelleher 2013). The sum of *proteoforms* of all proteins defines a specific state of a cell, tissue, or the whole organism. Proteomic approaches hence attempt to identify *proteoforms* differentially present in a diseased tissue.

7.3.1 Proteome Fractionation Techniques

Identification of the biomarkers is done by *mass spectrometry* (MS), the powerhorse of the proteomic discovery. Since MS technology has its limits as to the ability to resolve thousands of similar biomolecules, various protein separation and fractionation methods were designed to generate defined fractions of the proteomes to ease the identification in a mass spectrometer.

Liquid chromatography (LC) was the first of the modern separation techniques, which was developed in the early 1900s by the Russian botanist Mikhail S. Tswett. The method allows separation of large amounts of material based on proteins' hydrophobicity, size, charge properties, affinity to other proteins (e.g., antibodies), and presence of glycans or phosphogroups (Snyder et al. 2010). It is even easy to automate and standardize and allows sequential combination of two (even three) LC methods (reviewed in Dugo et al. (2008), Di Palma et al. (2012)). Its big advantage is a relatively straightforward interconnection to a mass spectrometer for the direct identification (or quantification) of the separated proteins or peptides.

Polyacrylamide gel electrophoresis in sodium dodecyl sulfate buffer (SDS-PAGE) is the second widely used protein separation technique. It has been invented by Laemmli in 1970 (Laemmli 1970). The SDS-denatured proteins are separated based on their size in a solid mesh of polymer filled with water and buffer molecules. The method was later updated to 2D electrophoresis, where proteins are first separated based on their charge and then (in the perpendicular direction) according to their size (O'Farrell 1975). The method was broadly applied to discovery proteomics (Oliveira et al. 2014).

Capillary zone electrophoresis (CE) is the newest from the classical proteomics techniques, although experimentation with this technique was documented already in 1930 by Tisselius (Petersen and Mohammad 2001). The proteins and peptides are separated in a thin capillary in an electric field based on their charge and size. The method is highly reproducible, fast, easily to automate, and highly sensitive with high resolution able to separate thousands of compounds (Stalmach et al. 2015). Furthermore, CE can be connected online to mass spectrometers, requires low sample volume, and is relatively cheap. The only disadvantage is the limited size of the proteins that can be effectively analyzed when coupled to a mass spectrometer (<20 kDa); therefore the protein mixture is usually digested with a protease and analyzed in the form of peptides (Pejchinovski et al. 2015).

Many other fractionation and enrichment techniques are used to simplify the complex proteomes, which include selective lysis of cells to preserve organelles and separate nuclei, membrane-associated proteins, and intra- and extracellular vesicles (synaptic vesicles, mitochondria, lysosomal and autosomal vesicles, exosomes from extracellular space, etc.) usually by differential centrifugation or isolation of protein complexes by using affinity resins (lectins, antibodies, etc.) (reviewed in Drissi et al. (2013)). Finally, in order to identify low abundant biomarkers or preserve those with limited half-life, it may be necessary to isolate the population of specific cells of interest, which may be performed by cell sorting in flow cytometry, by enrichment using immunomagnetic methods, and by microdissection (Altelaar and Heck (2012).

7.3.2 Mass Spectrometry

Mass spectrometry (MS) technology was initially developed by physicists for the measurement of the masses of atoms and lead, for example, to the discovery of isotopes (beginning of the twentieth century). In the last ~25 years, the biomarker discovery has been greatly facilitated by the quickly developing MS methods (the clinical applications were recently reviewed in Scherl 2015). This replaced the cumbersome methods of protein identification based on Edman N-terminal degradation, cloning, and antibody detection. Modern MS allows not only unequivocal identification of a protein but also its characterization including confirmation of the amino acid sequence and posttranslational modifications, i.e., identification of the individual *proteoforms*.

A mass spectrometer measures the mass to charge ratio (*m/z*) of an ionized molecule in an evacuated tube. Proteins, peptides, and organic molecules can be ionized by different methods, but two of them are particularly useful for proteomic application: electrospray (ESI) and matrix assisted laser desorption (MALDI). The *m/z* of the ions is then measured either in time-of-flight, quadrupole, ion trap, or ion cyclotron resonance mass analyzers (Scherl 2015).

MS is generally used for two applications: identification of unknown proteins and quantification of known targets. In the earlier stages of MS development, only small molecules could be effectively measured by mass spectrometers, and so the proteomes were first digested with a protease with defined cleavage sites (e.g., trypsin, LysC, LysN, etc.) (Giansanti et al. 2016) to shorter peptides amenable to MS analysis. This is called the "bottom-up" approach, and it still remains the mainstream of MS proteomics in various approaches.

In the *targeted proteomic* approach, the proteins can be quantified by selecting the protein-specific peptides (also posttranslationally modified) resulting from a digestion with a protease. Upon ionization, the peptide is selected and subjected to collision-induced fragmentation, and then the resulting selected fragment ion(s) is quantified. By introducing a defined amount of isotopically labeled internal standard for the protein, it is even possible to quantify its *absolute* amount. Furthermore, hundreds of peptides (proteins) can be quantified in the sample in parallel, thereby capturing a fairly complex dynamics of proteomes of the cell/tissue/organ (Soste et al. 2014). The targeted proteomics becomes a method of choice for the quantification of clinically important proteins and peptides, steadily replacing the widespread ELISA methods due to its high precision, reproducibility, flexibility (no need for labeling or antibodies), and multiplexing. Methods for MS quantification of amyloid β proteins and tau in CSF have been developed and implemented along with an MS-based method for characterization of reference material used for standardization of ELISA testing of CSF (Korecka et al. 2014; Leinenbach et al. 2014; McAvoy et al. 2014; Portelius et al. 2015; Barthelemy et al. 2016; Pannee et al. 2016). The drawbacks of MS quantification include slightly lower sensitivity and need of a state-of-the-art instrumentation.

The *shotgun* proteomics is the mainstream bottom-up approach. It enables identification of thousands of proteins in complex samples and is usually applied in discovery-based projects. In this approach all peptides in the digested proteomes are analyzed by a mass spectrometer. Depending on its speed, sensitivity, mass range, and resolution, the most abundant peptide ions are selected for fragmentation allowing the sequence confirmation and identification. The approach allows comparison of the protein/peptide abundance between two or more samples by labeling them with different isobaric tags, e.g., iTRAQ, SILAC, etc. (Rauniyar and Yates 2014), and thus identifies putative biomarkers. The abundance of peptides (and proteins) in the samples can even be compared by label-free quantification applying complex computational tools (Webb-Robertson et al. 2015).

The *bottom-up* approach is widely used for the analysis of posttranslational modifications of the proteins that could be associated with specific biological states of the cells or tissues. However, digestion of the proteins with a protease eliminates the possibility to compare the *proteoforms* of the expressed proteins between the two samples, because the specific pattern or combination of post-translational modifications is lost (Fig. 7.1), although some limited possibilities for analysis of proteoforms still exist (Lisitsa et al. 2014). Therefore, attempts are made to analyze the proteins in their intact forms, by the so-called *top-down* approach (Catherman et al. 2014).

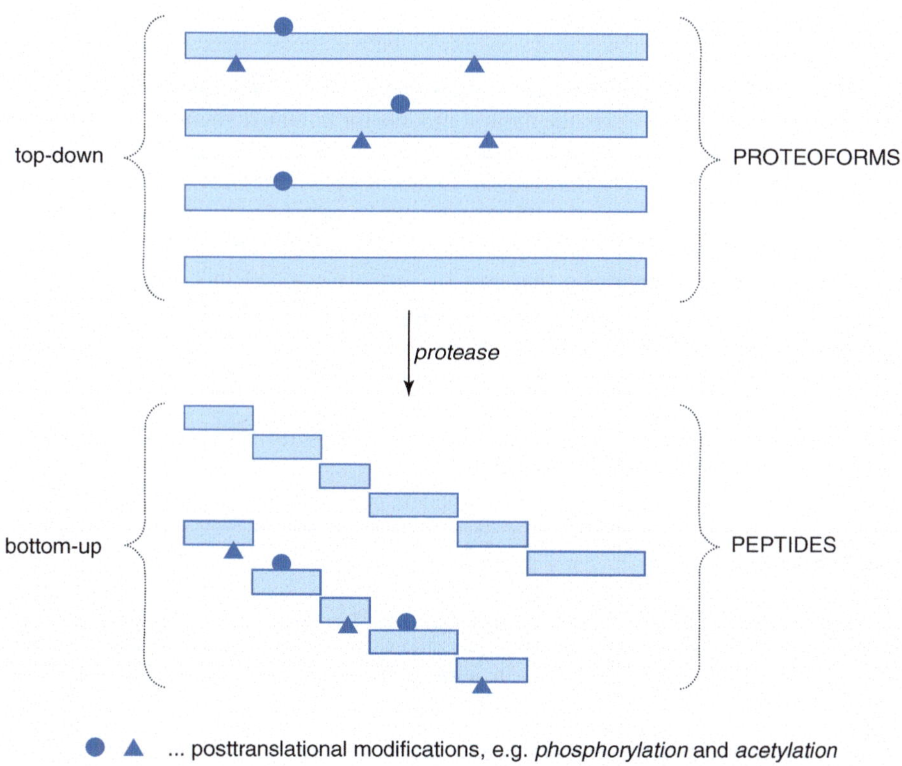

Fig. 7.1 Comparison of the top-down and bottom-up approaches in proteomics. Top-down approach preserves the proteoforms present in the sample, while digestion into peptides in the bottom-up approach degrades the proteoform information into a collection of subunits (peptides)

This approach poses a challenge for the MS instrumentation, since proteins larger than 20 kDa are difficult to analyze. When ionized by electrospray, the proteins attain multiple charges, which lead to "charge dilution" effect decreasing sensitivity. Furthermore, multiple charge forms of the proteins increase the chance that the ions of different proteins with the same m/z will overlap and hamper ion separation and characterization. To avoid such problems, the proteins need to be pre-fractionated and then the fractions introduced into a mass spectrometer. There are emerging techniques that solve these issues (Tran et al. 2011; Erba 2014; Molden and Garcia 2014; Sarsby et al. 2014; Scheffler 2014; Ye et al. 2014; Zhao et al. 2014; Guerrero et al. 2015) and allow label-free top-down quantitative comparison for biomarker development (Ntai et al. 2016).

The advances in proteomic methods and mass spectrometry instrumentation churning out systems with ever higher resolution, mass precision, robustness, and reproducibility move us slowly toward the complete characterization of the "protein complement" of the genome (Maurer et al. 2015; Wilson et al. 2015; Hu et al. 2016) and help us to understand physiological and pathological processes in the living organisms (Larance and Lamond 2015).

7.4 Summary

Biochemical markers related to the animal dementia are underdeveloped, despite the fact that dogs and cats provide natural models for human dementia. Biochemical diagnostics usually focuses on the body fluids as the most accessible sources of biological markers related to the disease. Cerebrospinal fluid and blood are the most interesting, since the former should contain the highest concentrations of the biomarkers and the latter is the easiest to collect. A number of immunoassays are in use for human dementia diagnosis that can be directly applied in canine and feline dementia diagnosis. Furthermore, dogs and cats provide excellent natural models of neurodegeneration and are therefore well suited for the identification of novel biomarkers using state-of-the-art proteomic methods.

Acknowledgment The authors received funding by APVV-0399-10, APVV-15-0613, and APVV-14-0872 and VEGA grants 2/0155/13 and 2/0164/16.

References

Adachi J, Kumar C, Zhang Y, Olsen JV, Mann M (2006) The human urinary proteome contains more than 1500 proteins, including a large proportion of membrane proteins. Genom Biol 7(9):1

Altelaar AM, Heck AJ (2012) Trends in ultrasensitive proteomics. Curr Opin Chem Biol 16(1):206–213

Anderson NL, Anderson NG (2002) The human plasma proteome: history, character, and diagnostic prospects. Mol Cell Proteom 2(1):50–50

Assarsson E, Lundberg M, Holmquist G, Bjorkesten J, Thorsen SB, Ekman D, Eriksson A, Rennel DE, Ohlsson S, Edfeldt G, Andersson AC, Lindstedt P, Stenvang J, Gullberg M, Fredriksson S (2014) Homogenous 96-plex PEA immunoassay exhibiting high sensitivity, specificity, and excellent scalability. PloS one 9(4):e95192

Barthelemy NR, Fenaille F, Hirtz C, Sergeant N, Schraen-Maschke S, Vialaret J, Buee L, Gabelle A, Junot C, Lehmann S, Becher F (2016) Tau protein quantification in human cerebrospinal fluid by targeted mass spectrometry at high sequence coverage provides insights into its primary structure heterogeneity. J Proteom Res 15(2):667–676

Beach TG, Monsell SE, Phillips LE, Kukull W (2012) Accuracy of the clinical diagnosis of Alzheimer disease at National Institute on Aging Alzheimer Disease Centers, 2005–2010. J Neuropathol Exp Neurol 71(4):266–273

Blennow K, Fredman P, Wallin A, Gottfries CG, Skoog I, Wikkelsö C, Svennerholm L (1993) Protein analysis in cerebrospinal fluid. II. Reference values derived from healthy individuals 18–88 years of age. Eur Neurol 33:129–133

Blennow K, Biscetti L, Eusebi P, Parnetti L (2016) Cerebrospinal fluid biomarkers in Alzheimer's and Parkinson's diseases—from pathophysiology to clinical practice. Movement Disord 31(6):836–847

Borghys H, Jacobs T, Van Broeck B, Dillen L, Dhuyvetter D, Gijsen H, Mercken M (2014) Comparison of two different methods for measurement of amyloid-β peptides in cerebrospinal fluid after BACE1 inhibition in a dog model. J Alzheimer's Dis 38(1):39–48

Bosch MN, Pugliese M, Gimeno-Bayon J, Rodriguez MJ, Mahy N (2012) Dogs with cognitive dysfunction syndrome: a natural model of Alzheimer's disease. Curr Alzheimer Res 9:298–314

Catherman AD, Skinner OS, Kelleher NL (2014) Top down proteomics: facts and perspectives. Biochem Biophys Res Commun 445(4):683–693

Chambers JK, Tokuda T, Uchida K, Ishii R, Tatebe H, Takahashi E, Tomiyama T, Une Y, Nakayama H (2015) The domestic cat as a natural animal model of Alzheimer's disease. Acta Neuropathol Commun 3:78

Chiam JTW, Dobson RJB, Kiddle SJ, Sattlecker M (2015) Are blood-based protein biomarkers for Alzheimer's disease also involved in other brain disorders? A systematic review. J Alzheimer's Dis 43(1):303–314

Chiu MJ, Yang SY, Chen TF, Chieh JJ, Huang TZ, Yip PK, Yang HC, Cheng TW, Chen YF, Hua MS, Horng HE (2012) New assay for old markers-plasma beta amyloid of mild cognitive impairment and Alzheimer's disease. Curr Alzheimer Res 9(10):1142–1148

Chiu MJ, Chen YF, Chen TF, Yang SY, Yang FP, Tseng TW, Chieh JJ, Chen JC, Tzen KY, Hua MS, Horng HE (2014) Plasma tau as a window to the brain—negative associations with brain volume and memory function in mild cognitive impairment and early alzheimer's disease. Human Brain Mapping 35(7):3132–3142

Devic I, Hwang H, Edgar JS, Izutsu K, Presland R, Pan C, Goodlett DR, Wang Y, Armaly J, Tumas V, Zabetian CP, Leverenz JB, Shi M, Zhang J (2011) Salivary alpha-synuclein and DJ-1: potential biomarkers for Parkinson's disease. Brain 134(7):e178

Di Palma S, Hennrich ML, Heck AJ, Mohammed S (2012) Recent advances in peptide separation by multidimensional liquid chromatography for proteome analysis. J Proteom 75(13): 3791–3813

Drissi R, Dubois ML, Boisvert FM (2013) Proteomics methods for subcellular proteome analysis. FEBS J 280(22):5626–5634

Dubois B, Feldman HH, Jacova C, Hampel H, Molinuevo JL, Blennow K, De Kosky ST, Gauthier S, Selkoe D, Bateman R, Cappa S, Crutch S, Engelborghs S, Frisoni GB, Fox NC, Galasko D, Habert MO, Jicha GA, Nordberg A, Pasquier F, Rabinovici G, Robert P, Rowe C, Salloway S, Sarazin M, Epelbaum S, de Souza LC, Vellas B, Visser PJ, Schneider L, Stern Y, Scheltens P, Cummings JL (2014) Advancing research diagnostic criteria for Alzheimer's disease: the IWG-2 criteria. Lancet Neurol 13:614–629

Dugo P, Cacciola F, Kumm T, Dugo G, Mondello L (2008) Comprehensive multidimensional liquid chromatography: theory and applications. J Chromatogr A 1184(1):353–368

Ellington AA, Kullo IJ, Bailey KR, Klee GG (2010) Antibody-based protein multiplex platforms: technical and operational challenges. Clin Chem 56(2):186–193

Erba EB (2014) Investigating macromolecular complexes using top-down mass spectrometry. Proteomics 14(10):1259–1270

Giansanti P, Tsiatsiani L, Low TY, Heck AJ (2016) Six alternative proteases for mass spectrometry-based proteomics beyond trypsin. Nat Protocols 11(5):993–1006

Gonzalez-Martinez A, Rosado B, Pesini P, Suarez ML, Santamarina G, Garcia-Belenguer S, Villegas A, Monleon I, Sarasa M (2011) Plasma beta-amyloid peptides in canine aging and cognitive dysfunction as a model of Alzheimer's disease. Exp Gerontol 46(7):590–596

Gren M, Shahim P, Lautner R, Wilson DH, Andreasson U, Norgren N, Blennow K, Zetterberg H (2016) Blood biomarkers indicate mild neuroaxonal injury and increased amyloid β production after transient hypoxia during breath-hold diving. Brain Injury 30(10):1226–1230

Guerrero A, Lerno L, Barile D, Lebrilla CB (2015) Top-down analysis of highly post-translationally modified peptides by Fourier transform ion cyclotron resonance mass spectrometry. J Am Soc Mass Spectrom 26(3):453–459

Guldbrandsen A, Vethe H, Farag Y, Oveland E, Garberg H, Berle M, Myhr KM, Opsahl JA, Barsnes H, Berven FS (2014) In-depth characterization of the cerebrospinal fluid (CSF) proteome displayed through the CSF proteome resource (CSF-PR). Mol Cell Proteom 13(11):3152–3163

Hu, A., Noble, W. S., Wolf-Yadlin, A. (2016). Technical advances in proteomics: new developments in data-independent acquisition. F1000 Research, 5

Hulstaert F, Blennow K, Ivanoiu A, Schoonderwaldt HC, Riemenschneider M, De Deyn PP, Bancher C, Cras P, Wiltfang J, Mehta PD, Iqbal K, Pottel H, Vanmechelen E, Vanderstichele H (1999) Improved discrimination of AD patients using β-amyloid (1-42) and tau levels in CSF. Neurology 52(8):1555–1562

Iliff JJ, Wang M, Liao Y, Plogg BA, Peng W, Gundersen GA, Benveniste H, Vates GE, Deane R, Goldman SA, Nagelhus EA, Nedergaard M (2012) A paravascular pathway facilitates CSF flow through the brain parenchyma and the clearance of interstitial solutes, including amyloid β. Sci Transl Med 4(147):147ra111

Johanson CE, Duncan JA, Klinge PM, Brinker T, Stopa EG, Silverberg GD (2008) Multiplicity of cerebrospinal fluid functions: new challenges in health and disease. Cerebrospinal Fluid Res 5(1)

Korecka M, Waligorska T, Figurski M, Toledo JB, Arnold SE, Grossman M, Trojanowski JQ, Shaw LM (2014) Qualification of a surrogate matrix-based absolute quantification method for amyloid-β42 in human cerebrospinal fluid using 2D UPLC-tandem mass spectrometry. J Alzheimer's Dis 41(2):441–451

Kovacech B, Novak M (2010) Tau truncation is a productive posttranslational modification of neurofibrillary degeneration in Alzheimer's disease. Curr Alzheimer Res 7(8):708–716

Laemmli UK (1970) Cleavage of structural proteins during the assembly of the head of bacteriophage T4. Nature 227:680–685

Larance M, Lamond AI (2015) Multidimensional proteomics for cell biology. Nat Rev Mol Cell Biol 16(5):269–280

Leinenbach A, Pannee J, Dulffer T, Huber A, Bittner T, Andreasson U, Gobom J, Zetterberg H, Kobold U, Portelius E, Blennow K (2014) Mass spectrometry-based candidate reference measurement procedure for quantification of amyloid-beta in cerebrospinal fluid. Clin Chem 60(7):987–994

Ling H, Hardy J, Zetterberg H (2015) Neurological consequences of traumatic brain injuries in sports. Mol Cell Neurosci 66:114–122

Lisitsa A, Moshkovskii S, Chernobrovkin A, Ponomarenko E, Archakov A (2014) Profiling proteoforms: promising follow-up of proteomics for biomarker discovery. Expert Rev Proteom 11(1):121–129

Lleo A, Cavedo E, Parnetti L, Vanderstichele H, Herukka SK, Andreasen N, Ghidoni R, Lewczuk P, Jeromin A, Winblad B, Tsolaki M, Mroczko B, Visser PJ, Santana I, Svenningsson P, Blennow K, Aarsland D, Molinuevo JL, Zetterberg H, Mollenhauer B (2015) Cerebrospinal fluid biomarkers in trials for Alzheimer and Parkinson diseases. Nat Rev Neurol 11(1):41–55

Madari A, Farbakova J, Katina S, Smolek T, Novak P, Weissova T, Zilka N (2015) Assessment of severity and progression of canine cognitive dysfunction syndrome using the CAnine DEmentia Scale (CADES). Appl Animal Behav Sci 171:138–145

Maurer MH (2010) Proteomics of brain extracellular fluid (ECF) and cerebrospinal fluid (CSF). Mass Spectrometry Rev 29(1):17–28

Maurer MM, Donohoe GC, Valentine SJ (2015) Advances in ion mobility-mass spectrometry instrumentation and techniques for characterizing structural heterogeneity. Analyst 140(20):6782–6798
McAvoy T, Lassman ME, Spellman DS, Ke Z, Howell BJ, Wong O, Zhu L, Tanen M, Struyk A, Laterza OF (2014) Quantification of tau in cerebrospinal fluid by immunoaffinity enrichment and tandem mass spectrometry. Clin Chem 60(4):683–689
McComb JG (1983) Recent research into the nature of cerebrospinal fluid formation and absorption. J Neurosurg 59(3):369–383
Molden RC, Garcia BA (2014) Middle-down and top-down mass spectrometric analysis of co-occurring histone modifications. Curr Protocols Protein Sci:23–27
Ntai I, Toby TK, LeDuc RD, Kelleher NL (2016) A method for label-free, differential top-down proteomics. Quantitat Proteom Mass Spectrometry 1410:121–133
O'Farrell PH (1975) High resolution two-dimensional electrophoresis of proteins. J Biol Chem 250(10):4007–4021
Oliveira BM, Coorssen JR, Martins-de-Souza D (2014) 2DE: the phoenix of proteomics. J Proteom 104:140–150
Olsson B, Lautner R, Andreasson U, Ohrfelt A, Portelius E, Bjerke M, Holtta M, Rosen C, Olsson C, Strobel G, Wu E, Dakin K, Petzold M, Blennow K, Zetterberg H (2016) CSF and blood biomarkers for the diagnosis of Alzheimer's disease: a systematic review and meta-analysis. Lancet Neurol 15(7):673–684
Pannee J, Gobom J, Shaw LM, Korecka M, Chambers EE, Lame M, Jenkins R, Mylott W, Carrillo MC, Zegers I, Zetterberg H, Blennow K, Portelius E (2016) Round robin test on quantification of amyloid-β 1–42 in cerebrospinal fluid by mass spectrometry. Alzheimer's Dementia 12(1):55–59
Pejchinovski M, Hrnjez D, Ramirez-Torres A, Bitsika V, Mermelekas G, Vlahou A, Zurbig P, Mischak H, Metzger J, Koeck T (2015) Capillary zone electrophoresis on-line coupled to mass spectrometry: a perspective application for clinical proteomics. Proteom Clin Appl 9(5-6):453–468
Petersen JR, Mohammad AA (2001) Clinical and forensic applications of capillary electrophoresis. Humana Press, New York
Portelius E, Lashley T, Westerlund A, Persson R, Fox NC, Blennow K, Revesz T, Zetterberg H (2015) Brain amyloid-beta fragment signatures in pathological ageing and Alzheimer's disease by hybrid immunoprecipitation mass spectrometry. Neurodegener Dis 15(1):50–57
Randall J, Mortberg E, Provuncher GK, Fournier DR, Duffy DC, Rubertsson S, Blennow K, Zetterberg H, Wilson DH (2013) Tau proteins in serum predict neurological outcome after hypoxic brain injury from cardiac arrest: results of a pilot study. Resuscitation 84(3):351–356
Rauniyar N, Yates JR III (2014) Isobaric labeling-based relative quantification in shotgun proteomics. J Proteome Res 13(12):5293–5309
Rissin DM, Kan CW, Campbell TG, Howes SC, Fournier DR, Song L, Piech T, Patel PP, Chang L, Rivnak AJ, Ferrell EP, Randall JD, Provuncher GK, Walt DR, Duffy DC (2010) Single-molecule enzyme-linked immunosorbent assay detects serum proteins at subfemtomolar concentrations. Nat Biotechnol 28(6):595–599
Rodríguez-Suárez E, Siwy J, Zürbig P, Mischak H (2014) Urine as a source for clinical proteome analysis: from discovery to clinical application. Biochim Biophys Acta (BBA)-Proteins Proteomics 1844(5):884–898
Roerig A, Carlson R, Tipold A, Stein VM (2013) Cerebrospinal fluid tau protein as a biomarker for severity of spinal cord injury in dogs with intervertebral disc herniation. Veterinary J 197(2):253–258
Sarsby J, Martin NJ, Lalor PF, Bunch J, Cooper HJ (2014) Top-down and bottom-up identification of proteins by liquid extraction surface analysis mass spectrometry of healthy and diseased human liver tissue. J Am Soc Mass Spectrom 25(11):1953–1961
Scheffler K (2014) Top-down proteomics by means of Orbitrap mass spectrometry. Shotgun Proteomics: Methods and Protocols, 465–487
Scherl A (2015) Clinical protein mass spectrometry. Methods 81:3–14

Schütt T, Toft N, Berendt M (2015) Cognitive function, progression of age-related behavioral changes, biomarkers, and survival in dogs more than 8 years old. J Veterinary Intern Med 29(6):1569–1577

Schutzer SE, Liu T, Natelson BH, Angel TE, Schepmoes AA, Purvine SO, Hixson KK, Lipton MS, Camp DG, Coyle PK, Smith RD, Bergquist J (2010) Establishing the proteome of normal human cerebrospinal fluid. PLoS One 5(6):e10980

Shahim P, Tegner Y, Wilson DH, Randall J, Skillback T, Pazooki D, Kallberg B, Blennow K, Zetterberg H (2014) Blood biomarkers for brain injury in concussed professional ice hockey players. JAMA Neurol 71(6):684–692

Shi M, Sui YT, Peskind ER, Li G, Hwang H, Devic I, Ginghina C, Edgar JS, Pan C, Goodlett DR, Furay AR, Gonzalez-Cuyar LF, Zhang J (2011) Salivary tau species are potential biomarkers of Alzheimer's disease. J Alzheimer's Dis 27(2):299–305

Smith LM, Kelleher NL (2013) Proteoform: a single term describing protein complexity. Nat Meth 10(3):186–187

Snyder LR, Kirkland JJ, Dolan JW (2010) Introduction to modern liquid chromatography. Wiley, New York

Soste M, Hrabakova R, Wanka S, Melnik A, Boersema P, Maiolica A, Wernas T, Tognetti M, von Mering C, Picotti P (2014) A sentinel protein assay for simultaneously quantifying cellular processes. Nat Meth 11(10):1045–1048

Stalmach A, Husi H, Mosbahi K, Albalat A, Mullen W, Mischak H (2015) Methods in capillary electrophoresis coupled to mass spectrometry for the identification of clinical proteomic/peptidomic biomarkers in biofluids. Meth Mol Biol 1243:187–205

Tarasoff-Conway JM, Carare RO, Osorio RS, Glodzik L, Butler T, Fieremans E, Axel L, Rusinek H, Nicholson C, Zlokovic BV, Frangione B, Blennow K, Menard J, Zetterberg H, Wisniewski T, de Leon MJ (2015) Clearance systems in the brain [mdash] implications for Alzheimer disease. Nat Rev Neurol 11(8):457–470

Tarawneh R, Lee JM, Ladenson JH, Morris JC, Holtzman DM (2012) CSF VILIP-1 predicts rates of cognitive decline in early Alzheimer disease. Neurology 78(10):709–719

Tarawneh R, Head D, Allison S, Buckles V, Fagan AM, Ladenson JH, Morris JC, Holtzman DM (2015) Cerebrospinal fluid markers of neurodegeneration and rates of brain atrophy in early Alzheimer disease. JAMA Neurol 72(6):656–665

Tarawneh R, D'Angelo G, Crimmins D, Herries E, Griest T, Fagan AM, Zipfel GJ, Ladenson JH, Morris JC, Holtzman DM (2016) Diagnostic and prognostic utility of the synaptic marker neurogranin in Alzheimer disease. JAMA Neurol 73(5):561–571

Tran JC, Zamdborg L, Ahlf DR, Lee JE, Catherman AD, Durbin KR, Tipton JD, Vellaichamy A, Kellie JF, Li M, Wu C, Sweet SM, Early BP, Siuti N, LeDuc RD, Compton PD, Thomas PM, Kelleher NL (2011) Mapping intact protein isoforms in discovery mode using top-down proteomics. Nature 480(7376):254–258

Vandermeeren M, Mercken M, Vanmechelen E, Six J, Voorde A, Martin JJ, Cras P (1993) Detection of proteins in normal and Alzheimer's disease cerebrospinal fluid with a sensitive sandwich enzyme-linked immunosorbent assay. J Neurochem 61(5):1828–1834

Webb-Robertson BJ, Wiberg HK, Matzke MM, Brown JN, Wang J, McDermott JE, Smith RD, Rodland KD, Metz TO, Pounds JG, Waters KM (2015) Review, evaluation, and discussion of the challenges of missing value imputation for mass spectrometry-based label-free global proteomics. J Proteome Res 14(5):1993–2001

Werle M, Bernkop-Schnürch A (2006) Strategies to improve plasma half life time of peptide and protein drugs. Amino Acids 30(4):351–367

Wilson SR, Vehus T, Berg HS, Lundanes E (2015) Nano-LC in proteomics: recent advances and approaches. Bioanalysis 7(14):1799–1815

Yang T, O'Malley TT, Kanmert D, Jerecic J, Zieske LR, Zetterberg H, Hyman BT, Walsh DM, Selkoe DJ (2015) A highly sensitive novel immunoassay specifically detects low levels of soluble Aβ oligomers in human cerebrospinal fluid. Alzheimer's Res Therapy 7:14

Ye H, Mandal R, Catherman A, Thomas PM, Kelleher NL, Ikonomidou C, Li L (2014) Top-down proteomics with mass spectrometry imaging: a pilot study towards discovery of biomarkers for neurodevelopmental disorders. PloS one 9(4):e92831

Zafari S, Backes C, Meese E, Keller A (2015) Circulating biomarker panels in Alzheimer's disease. Gerontology 61(6):497–503

Zetterberg H, Mortberg E, Song L, Chang L, Provuncher GK, Patel PP, Ferrell E, Fournier DR, Kan CW, Campbell TG, Meyer R, Rivnak AJ, Pink BA, Minnehan KA, Piech T, Rissin DM, Duffy DC, Rubertsson S, Wilson DH, Blennow K (2011) Hypoxia due to cardiac arrest induces a time-dependent increase in serum amyloid β levels in humans. PloS one 6(12):e28263

Zetterberg H, Wilson D, Andreasson U, Minthon L, Blennow K, Randall J, Hansson O (2013) Plasma tau levels in Alzheimer's disease. Alzheimer's Res Ther 5:9

Zhao Y, Sun L, Champion MM, Knierman MD, Dovichi NJ (2014) Capillary zone electrophoresis–electrospray ionization-tandem mass spectrometry for top-down characterization of the Mycobacterium marinum secretome. Analyt Chem 86(10):4873–4878

Zilka N, Kovacech B, Barath P, Kontsekova E, Novak M (2012) The self-perpetuating tau truncation circle. Biochem Soc Trans 40:681–686

Current Pharmacological and Non-pharmacological Approaches for Therapy of Feline and Canine Dementia

8

Sagi Denenberg and Gary Landsberg

Treatment options for cognitive dysfunction syndrome (CDS) could target prevention, slowing decline and improving clinical signs. However, while prompt diagnosis and early initiation of treatment have the greatest potential for success, most owners do not report signs until they are significantly advanced and veterinarians fail to educate owners about the importance of early reporting. Both laboratory trials in beagle dogs and in cats using neuropsychological assessment tools to assess learning and memory and clinical trials have been used to assess effect of therapeutics.

There are four treatment modalities that are presently available which can be used alone or in combination that target improving brain metabolism, enhancing neuronal transmission, reducing oxidative damage, and helping to maintain neuronal integrity. These include drugs, natural supplements, therapeutic diets, and environmental enrichment. In addition new research is ongoing into vaccination therapy and stem cell therapy for cognitive impairment. Drugs marketed for treatment of cognitive dysfunction and mental confusion in senior dogs include selegiline, propentofylline, and nicergoline. While no drugs have been licensed for cats, there is anecdotal evidence of improvement with these canine medications. Supplements include *S*-adenosylmethionine and combinations of ingredients that might include phosphatidylserine, resveratrol, vitamins, and fatty acids. Dietary therapeutics include a senior diet for canine cognitive dysfunction enriched with antioxidants,

S. Denenberg
Langford Veterinary Services Ltd., University of Bristol, Bristol, UK
e-mail: sagidvm@gmail.com

G. Landsberg (✉)
North Toronto Veterinary Behaviour Specialty Clinic, Thornhill, ON, Canada

CanCog Technologies, Toronto, ON, Canada
e-mail: gmlandvm@aol.com

G. Landsberg et al. (eds.), *Canine and Feline Dementia*,
DOI 10.1007/978-3-319-53219-6_8

mitochondrial cofactors such as alpha-lipoic acid and L-carnitine, and fatty acids and another which is supplemented with medium-chain triglycerides (MCT) to provide ketones as an alternative source of energy for the aging brain. Similar diets are not yet available for cats, although there have been studies demonstrating potential effects of dietary therapeutics for cats. Finally, studies have shown that behavioral and environmental enrichment can aid in slowing decline.

8.1 Treatment of CDS

Treatment of cognitive dysfunction and its clinical signs should be focused on three different but closely related targets: to improve the clinical signs of cognitive dysfunction for the welfare and well-being of the pet and the owner, to slow the progression of decline, and to prevent or delay the onset of dementia. The treatment options presently available are diet, supplements, and drugs. In addition, studies in dogs have demonstrated that enrichment in the form of mental stimulation and physical exercise can improve quality of life and delay the onset or slow the decline of both the behavioral signs and pathology associated with brain aging, particularly if combined with an appropriate nutritional base (Head et al. 2009; Milgram et al. 2004; Head 2007; Valenzuela and Sachdev 2009; Siwak-Tapp et al. 2008). This is consistent with human studies in which higher education, mental enrichment, and physical exercise and nutrition have been found to delay the onset of dementia (Geda et al. 2010; Scarmeas et al. 2006; Joseph et al. 2009; Donini et al. 2007; Kidd 2005).

To date, success in treating cognitive decline has been limited, in part due to the delay in making a diagnosis and in beginning treatment too late in the course of the decline. In many cases, signs of CDS are present for an extended period of time before owners seek veterinary advice. (See Chap. 1 on clinical signs of cognitive dysfunction). In addition, since CDS is a diagnosis of exclusion in both dogs and cats, and health issues can arise concurrent with CDS, treatment is further delayed until other possible causes of the clinical signs are diagnosed and treated (see Chap. 2 for behavioral and medical differentials). Prompt diagnosis and early initiation of treatment should achieve the greatest success. Therefore, it is paramount that the veterinarian screen senior and geriatric pets ideally twice a year and utilize questionnaires to efficiently insure effective and comprehensive screening. Owner education, including handouts and web resources provided to the pet owner beginning at middle age, will increase awareness of normal vs. abnormal aging and educate owners as to the importance of promptly reporting any behavioral signs.

8.2 Pharmacological Therapy

While most therapeutic options available for treatment of CDS are labeled for use in dogs, a few are also labeled for use in cats, although there is minimal evidence to support efficacy in cats. Thus, the potential benefits must be weighed against possible risks.

8.2.1 Selegiline

In North America, selegiline hydrochloride (Anipryl®, Zoetis Animal Health) is the only pharmaceutical that is approved for the treatment of CDS in aged dogs. Selegiline has demonstrated an improvement of the clinical signs of CDS and an improvement in working memory in a laboratory model at a dose of 0.5–1 mg/kg in the morning (Ruehl et al. 1995; Head et al. 1996; Campbell et al. 2001).

Selegiline is categorized as a selective irreversible inhibitor of monoamine oxidase B (MAOB) although its mode of action in dogs is not entirely clear. Selegiline has been shown to increase 2-phenylethylamine in the cortex and hippocampus, a neuromodulator that enhances dopamine and catecholamine function. Selegiline may also alleviate CDS by increased release and decreased reuptake of norepinephrine. Its metabolites L-amphetamine and L-methamphetamine may further enhance cognitive function and improve behavior. Selegiline may also have neuroprotective effects on dopaminergic, noradrenergic, and cholinergic neurons (Milgram et al. 1993; Milgram et al. 1995; Ruehl et al. 1996).

Selegiline may also contribute to a decrease in the production and increase in the clearance of free radical load in the brain, by increasing scavenging enzymes such as superoxide dismutase and catalase (Carrillo et al. 1994). Some dogs improve within the first 2 weeks, while a few do not show improvement until the second month (Campbell et al. 2001).

Although feline use is off label, beneficial effects in cats with signs of CDS including disorientation, vocalization, and decreased interest in affection have been reported at a dose of 0.5–1 mg/kg per day in the morning (Landsberg et al. 2010; Gunn-Moore et al. 2007; Overall 2010; Landsberg 2006). In one small study, no toxicity was seen in a dose up to 10 mg/kg (Ruehl et al. 1996).

Selegiline should not be used concurrently with other MAO inhibitors such as amitraz and drugs that might increase serotonin transmission, such as SSRIs and tricyclic antidepressants, tramadol, buspirone, and most narcotics. A withdrawal time of at least two weeks is suggested between withdrawal of selegiline and initiating SSRI or TCA therapy and ideally a washout of 5 weeks between the discontinuation of fluoxetine and the start of selegiline although a shorter washout e.g. two weeks should be sufficient with cautious.

8.3 Propentofylline

Propentofylline is a xanthine derivative which may improve microcirculation and inhibit platelet aggregation and thrombus formation to increase oxygenation to the brain and periphery. It is licensed for use in dogs in some countries in Europe and in Australia (Vivitonin®, Karsivan Vet®, MSD Animal Health) at 5 mg/kg twice daily for signs of senility including mental dullness, lethargy, and tiredness when no other underlying medical cause is identified (www.msd-animal-health.co.nz; Parkinson et al. 1994). In a laboratory trial with aged beagle dogs, it had no effect on behavioral activity (Siwak et al. 2000). In cats, there are anecdotal reports of efficacy at 12.5 mg/cat orally twice daily (Landsberg et al. 2010; Gunn-Moore et al. 2007; Overall 2010).

8.4 Cholinergic Decline

Elderly pets have a decline in cholinergic function, and anticholinergic medications further aggravate cognitive decline (Zhang et al. 2005; Pugliese et al. 2007; Araujo et al. 2005a, 2011a, b). In fact, use of anticholinergic drugs might potentially contribute to further cognitive impairment (Cai et al. 2013; Gray et al. 2015; Cross et al. 2016). Therefore, it is prudent to avoid anticholinergic medications in senior pets (Araujo et al. 2005a, 2011a, b). Moreover, drugs or natural products that enhance cholinergic transmission might have potential benefits for improving signs of CDS in dogs and cats, but more research is required to find safe and effective drugs and doses (Araujo et al. 2011a). For a list of drugs with anticholinergic effects, see http://www.agingbraincare.org/uploads/products/ACB_scale_-_legal_size.pdf.

Other treatment strategies include nicergoline an alpha 1 and alpha 2 agonist (given at 0.5 mg/kg sid), which may act to enhance cerebral circulation and enhance neuronal transmission and may have a neuroprotective effect (Siwak et al. 2000; Penaliggon 1997). However, in one study both nicergoline and propentofylline resulted in no significant increase in locomotion in dogs compared to adrafinil and modafinil, which might enhance noradrenergic transmission. Therefore, these medications might be a consideration to increase mental alertness and daytime activity. However, while adrafinil was demonstrated to improve discrimination learning, it led to impaired memory (Siwak et al. 2003).

8.5 Natural Supplements

In addition to pharmacological intervention, natural supplements may be used. Supplements have potentially less side effects and are not generally contraindicated with most drugs or disease processes (e.g., renal, hepatic, or cardiac dysfunction).

Senilife® (Ceva Animal Health) is a product labeled for both dogs and cats that contains phosphatidylserine, an important building block of cell membranes that is purported to facilitate neuronal signal transduction and enhance cholinergic transmission (Tsakiris and Deconstantinos 1985; Osella et al. 2007). Senilife has been demonstrated to improve cognition in both a laboratory model and clinical studies in dogs (Osella et al. 2007; Araujo et al. 2008). In addition to phosphatidylserine, the supplement contains *Ginkgo biloba*, vitamin E, and resveratrol for their potential antioxidant effects and also contains vitamin B6 (pyridoxine) which may have neuroprotective effect (Dakshinamurti et al. 2003). No efficacy data has been published for cats.

Aktivait® (VetPlus Ltd.) which contains phosphatidylserine®, omega-3 fatty acids, vitamins E and C, L-carnitine, alpha-lipoic acid, coenzyme Q, and selenium has also demonstrated significant improvement over placebo in improving social interactions, disorientation, and house soiling in dogs with CDS (Heath et al. 2007). The feline version of the product does not contain alpha-lipoic acid because of potential for toxicity; however, efficacy has not yet been evaluated in controlled trials (Hill et al. 2004).

S-adenosylmethionine (Novifit®, Virbac Animal Health) is a natural supplement for dogs and cats that may help to maintain cell membrane fluidity and receptor function, regulate neurotransmitter levels, and increase production of glutathione which may decrease oxidative stress (Rème et al. 2008; Araujo et al. 2012). Improvement in clinical signs of cognitive dysfunction in dogs has been demonstrated in a placebo-controlled trial (Rème et al. 2008). In laboratory testing in aged beagle dogs, there was a trend toward improved reversal learning. In reversal learning in aged cats, there was a nonsignificant reduction in errors in the treatment group compared to the control group. However, when the cats were divided into top and bottom half performers, the top performers had significantly less reversal learning errors consistent with improved executive function. This suggests that Novifit improves an age-related decline in executive function that is less beneficial in cats that are most affected (Araujo et al. 2012).

Another supplement for senior cats, Cholodin®-Fel (MVP Labs), contains choline, phosphatidylcholine, methionine, inositol, vitamin E, zinc, selenium, taurine, and vitamin B. In one preliminary study, 9 of 21 aged cats showed improvement in confusion and appetite (Messonier 2001).

Apoaequorin (Neutricks™, Neutricks LLC) is a protein found in jellyfish that in two separate laboratory trials improved learning and attention in dogs. It is a calcium-buffering protein that may provide neuroprotection against aging (Milgram et al. 2015). Dysregulation of intracellular calcium is associated with increased age and may be linked to cognitive dysfunction syndrome in dogs. In one study, the apoaequorin-treated animals showed improved performance on discrimination learning and attention tasks but did not differ from the controls on the spatial memory task, while in a second study, apoaequorin showed superior performance to selegiline. A cat product is available, but no evidence-based studies have been published.

Other natural products might be considered as cognitive supplements based on purported but inconclusive evidence of effect on humans with Alzheimer's disease. These include Huperzine A, derived from Chinese club moss (*Huperzia serrata*) which is licensed for treatment of Alzheimer's disease in China. It acts as a cholinesterase inhibitor as well as having possible neuroprotective effects against oxidative injury and as an NMDA antagonist (Qian and Ke 2014; Yang et al. 2013). Curcurmin, an antioxidant, anti-amyloid, and anti-inflammatory compound found in the turmeric and catechin spices, is postulated to be helpful (Yang et al. 2013).

8.6 Nutritional Intervention

A primary therapeutic strategy for cognitive dysfunction in dogs, cats, and humans is to reduce the risk factors that contribute to cognitive decline through nutritional intervention with a goal of maintaining brain health, slowing decline, and improving clinical signs and pet welfare. Strategies for treatment include individual supplements and combinations or "cocktails" of ingredients, focused on reducing the effects of oxidative stress, correcting metabolic changes associated with cognitive decline, and improving mitochondrial function, neuronal health, and signaling (Head 2007; Head et al. 2002, 2009; Milgram et al. 2004; Valenzuela and Sachdev 2009; Siwak-Tapp et al. 2008; Kidd 2005; Landsberg et al. 2010; Gunn-Moore et al. 2007; Sullivan and Brown 2005; Pan et al. 2010, 2013, 2017; Pan 2011; Scarmeas et al. 2006; Joseph et al. 2009; Araujo et al. 2005 b; Dowling and Head 2012; Fahnestock et al. 2012; Pop et al. 2010). In addition, in a recent study of dogs with cognitive dysfunction syndrome in Slovakia, when comparing controlled diets (quality commercial foods for size, breed, age, or health status) to uncontrolled diets (low-quality commercial foods or leftovers), risk for cognitive dysfunction was increased with the lower-quality diets (Katina et al. 2016).

(a) Therapeutic diets

The first diet studied for the treatment of cognitive dysfunction was formulated by Hill's Pet Nutrition (Prescription Diet b/d Canine) with a focus on improving antioxidant defenses and reducing the effects of oxidative damage. It is supplemented with fatty acids, antioxidants (vitamins C and E, beta-carotene, selenium, flavonoids, carotenoids), and dl-alpha-lipoic diet and L-carnitine which are intended to enhance mitochondrial function. The diet was evaluated in 48 beagle dogs aged 8–12 in a longitudinal study over 3 years on a variety of cognitive tasks, with improvement in a landmark task within 2 weeks and subsequent improvement in a more complex task (oddity discrimination). Visual discrimination improved, and reversal learning was maintained over time while the control group declined. Four different groups were assessed: (a) added behavioral enrichment consisting of social enrichment, physical exercise, and cognitive training plus antioxidant diet, (b) behavioral enrichment with control diet, (c) no added behavioral enrichment with antioxidant diet, and (d) no behavioral enrichment plus control diet. The highest cognitive scores were seen in the dogs that received both the antioxidant diet and the added enrichment with spatial memory and reversal learning improving over 2 years. When the analysis took into consideration the enrichment as well as the supplemented diet, the greatest effect on improving and maintaining cognitive function was in the group receiving both antioxidant supplement diet and enrichment for both reversal and memory tasks (Head et al. 2009; Milgram et al. 2004, Head 2007; Araujo et al. 2005 b; Dowling and Head 2012; Fahnestock et al. 2012). By contrast, there was no improvement in young dogs fed with the cognitive diet (Scarmeas et al. 2006). The combined effect of the diet and enrichment acted, perhaps in a synergistic manner, to reduce oxidative damage and increase brain-derived neurotropic factor and neuronal health (Head et al. 2009; Milgram et al. 2004; Head 2007; Dowling and Head 2012; Fahnestock et al. 2012; Pop et al. 2010). While enrichment protected against neuronal loss in the hippocampus (Siwak-Tapp et al. 2008; Dowling and Head 2012), mitochondrial function and beta-amyloid pathology were significantly improved with diet and not with enrichment alone (Head et al. 2009; Dowling and Head 2012; Pop et al. 2010). However, a reduction in beta-amyloid did not correlate with cognitive performance (Pop et al. 2010). Interestingly, in the supplementation with alpha-lipoic acid, L-carnitine, or both, no cognitive benefits were seen (Christie et al. 2009). In a clinical trial, significant effects were obtained from the diet alone (Dodd et al. 2003).

A diet from Nestle Purina Research (Purina Pro Plan Bright Mind (North America), Optiage (UK)) is supplemented with botanic oils containing medium-chain triglycerides (MCT) to provide ketone bodies that might serve as an alternative source of energy for aging neurons. In fact, studies have shown a significant reduction in cerebral glucose metabolism by 6 years of age compared to dogs at one year of age (London et al. 1983). Over an 8-month trial, the group supplemented with 5.5% MCT showed significantly better performance over a placebo diet in a variety of test protocols including landmark 1 and landmark 2 discriminations, an egocentric learning and egocentric reversal task, and a variable oddity task (Pan et al, 2010; Pan 2011). The group given MCT supplement showed significantly elevated levels of the ketone body, β-hydroxybutyrate (Sullivan and Brown 2005; Taha et al. 2009). Most recently in a double-blinded placebo-controlled clinical trial in dogs with CDS, a diet supplemented with 6.5% MCT and a brain protection blend (BPB) described below demonstrated significant improvement in all 6 categories of DISHAA (see table 1.1) over a 3-month trial (ProPlan Veterinary Diets Neurocare) (Pan et al, 2017).

Royal Canin Canine Mature Consult and Feline Mature Consult diets are also formulated with a blend of phosphatidylserine, antioxidants, and L-tryptophan to help maintain cognitive health in senior pets.

One study by Nestle Purina utilized a proprietary brain protection blend (BPB) containing B vitamins, antioxidants including vitamins E and C and selenium, fish oil containing DHA and EPA for anti-inflammatory effects, and arginine to enhance nitric oxide synthesis to reduce blood pressure and improve circulation and cognition (Pan et al. 2013; Dong et al. 2011; Selhub et al. 2010). In a group of 32 cats aged 5.5 to 8.7, over 1 year the BPB diet showed better performance than the control diet on egocentric learning and reversal, discrimination learning, and acquisition of a spatial memory task (Pan et al. 2013). Another study by Hill's Pet Nutrition evaluated a test diet supplemented with tocopherols, L-carnitine, vitamin C, beta-carotene, DHA, cysteine, and methionine designed to decrease production and increase clearance of free radicals over 30 days in 44 cats, 10 years and over. The cats fed with the test diet had significantly improved activity, compared to when the cat was 8 years of age and compared to the control group (Houpt et al. 2007). These latter two feline diets are not commercially available.

8.7 New Horizons

Immunotherapy has also been evaluated in aged canines involving the vaccination of aged animals against the Aβ protein (Head et al. 2008; Bosch et al. 2013). Although the study was not successful in reversing cognitive deficits, this approach may have future treatment potential (Head et al. 2008). In fact, a recent study found that the combined effects of vaccination and behavioral enrichment resulted in maintenance of learning, increased BDNF mRNA, and a decline in beta-amyloid (Davis et al. 2016). Another strategy under active investigation, with anecdotal reports of efficacy in improving signs of cognitive dysfunction, is the use of stem cell therapy obtained from either the skin and implanted into the hippocampus (Valenzuela et al. 2008) or from the olfactory mucosa and implanted into the cisterna magna (Veron et al. 2014).

8.7.1 Adjunctive Therapy

Together with diets, supplements, and drugs for the treatment for CDS, psychotropic medications may be required concurrently to manage underlying stress and address those signs such as night waking, agitation, and anxiety that continue to be problematic for the owner and pet (Table 8.1). As the health and behavior of senior pets may necessitate the use of multiple medications, caution must be exercised to insure that there are no incompatibilities or contraindications when combining medications and supplements. For example, selegiline should not be used in combination with MAO inhibitors such as amitraz or drugs that might increase serotonin.

Since anticholinergic drugs should be avoided, fluoxetine, sertraline, or buspirone might be considered as ongoing therapeutics and clomipramine, amitriptyline, and paroxetine avoided (Zhang et al. 2005; Pugliese et al. 2007; Araujo et al. 2005a, 2011a, b). Trazodone might also be considered either alone or in combination with an ongoing SSRI or buspirone. While benzodiazepines could contribute to further cognitive deficits, sedation, or incoordination, they may be useful in managing signs of anxiety and sleep disturbances. Lorazepam and oxazepam would be preferred since they have no active intermediate metabolites. Adjunctive use of propranolol or clonidine or dexmedetomidine oromucosal gel may inhibit release or block the effects of noradrenaline. Gabapentin might reduce reactivity and neuropathic pain (Landsberg et al. 2009, 2011). Natural products might also aid in the control of anxiety including pheromones (Adaptil, Feliway), L-theanine, alpha-casozepine, melatonin, or supplements containing GABA, phellodendron and magnolia, or souroubea.

Table 8.1 Drug doses for adjunctive therapy of clinical signs

	Dog dose	Cat dose
Lorazepam	.025–0.1 mg/kg sid to prn	.125–.25 mg/cat sid-bid
Oxazepam	0.2–1.0 mg/kg sid-bid	0.2–0.5 mg/kg sid-bid
Fluoxetine	1.0–2.0 mg/kg sid	0.5–1.5 mg/kg sid
Sertraline	1–5 mg/kg sid or divided bid	0.5–1.5 mg/kg sid
Clonidine	0.01–0.05 mg/kg bid or prn	5–10 micrograms/kg bid-tid
Dexmedetomidine oromucosal gel	125 μgrams/m^2	
Propranolol	0.5–2.0 mg/kg bid or prn	
Buspirone	0.5–2.0 mg/kg bid-tid	.5–1 mg/kg bid
Trazodone	2–5 mg/kg (to 12 mg/kg) prn to tid	25–50 mg/cat sid-bid
Gabapentin	10–30 mg/kg bid-tid	5–20 mg/kg sid-tid
Selegiline	0.5–1 mg/kg sid in am	0.5–1 mg/kg sid in am

8.8 Drug Doses

Doses of drugs used in veterinary behavior as adjunctive therapy for signs related to anxiety, agitation, and irritability, night waking, and soiling are listed below. For senior pets, for the safest and most prudent use of drugs, start at lowest suggested dose and increase gradually if there is insufficient therapeutic effect and no adverse effects. As discussed, drugs with anticholinergic effects should be avoided where possible, and drugs that might sedate, contribute to ataxia or incoordination, or have an impact on cognition should be used cautiously or avoided.

8.8.1 Behavioral and Environmental Management

Together with medical and nutritional therapy, providing both physical stimulation and mental enrichment are important aspects of maintaining behavioral and physical health and slowing decline (Head et al. 2009; Milgram et al. 2004; Head 2007). However, since senior pets may develop an increasing number of medical health issues with increasing age including sensory decline (sight and hearing), endocrinopathies (e.g., Cushing's, diabetes, and hyperthyroidism), organ dysfunction (e.g., renal and hepatic), increasing pain (e.g., osteoarthritis), and decline in mobility, the amount, level, and type of enrichment will need to be tailored to the needs and limitations of the individual.

Meeting the individual needs of the pet is the primary consideration when designing an optimal enrichment plan. These needs include feeding, sleeping, grooming and self-hygiene, elimination, and attention (affection and interaction) from the owner. In addition, exercise and play should be considered as well. For cats, climbing, perching, and scratching opportunities should be provided, while dogs (and some cats) should have outlets to chew. Therefore, the veterinarian and owner should carefully assess the pet's needs, abilities, and limitations when establishing an enrichment plan.

Behavioral enrichment should include social interactions with the owner such as play, scent work, and training. While training might take more time, patience, and persistence, in senior pets, it is possible. Dogs and cats can be trained to find food hidden in a room or yard, follow basic cues, and perform tricks. Clicker and target training can be incorporated into training.

The environment should be enriched as well for exploring, playing, and scavenging. Owners should continue to offer existing toys and add additional novel toys to senior pets. Owners may need to initiate play and reward the pet for taking part. The use of feeding toys as part of enrichment is simple and generally quite successful. Puzzle toys, maze toys, and other food-dispensing toys (e.g., Buster DogMaze, Kong Wobbler, Cat Activity Fun Board, or Kitty Kong) can be used for regular feeding or treats. These can be left when the owner is absent providing the pet desired and pleasant activity.

Providing senior pets with a specific location (safe haven) at home where they are encouraged to spend quality time (including resting, chewing, and object play) can help in reducing anxiety and confusion. Owners should select a location that is easily accessible by the pet, place bedding, feeding toys, and water there and consider a pheromone diffuser. The pet should be taken there several times daily, for rest, food, affection, and play. This place should be associated only with positive experience and will provide the pet with comfort and calmness.

Behavioral and emotional stability in senior pets is achieved via owners' positive interactions, predictability, and structured daily routine. Owners are encouraged to have a daily schedule for their pet that includes all its needs throughout the day. This way the pet knows what to expect, regulate bodily function and metabolism, and have its biological rhythm adjusted. All changes (move, new family members, new pet, etc.) should be introduced gradually and coupled with positive reinforcement. All these will help to minimize fear, anxiety, and stress, which can further contribute to a decline in mental and physical health.

8.9 Considerations for Senior Pet Enrichment

Canine: Social activities and exercise (should be selected or modified taking into consideration pet's physical, mental, and social limitations):

(a) Physical activities including walks or runs, swimming, retrieve, tug, chase, or frisbee and working activities or sports such as herding, hunting, or agility
(b) Intraspecific play—other family pets, dog parks
(c) Reward-based training—practicing learned behaviors/learning and shaping new behaviors

Exploratory play:

(a) Food puzzle and manipulation toys
(b) Food search, learning toys, and nose work/scent training
(c) Chews and chew toys

Resources: https://positively.com/dog-wellness/dog-enrichment; www.indoorpet.osu.edu

8.10 Feline

Social activities and exercise (should be selected or modified taking into consideration the pet's physical, mental and social limitations):

(a) Physical activities including playing with multiple chase toys as outlets for predation and walking on harness
(b) Intraspecific play with other pets where applicable
(c) Reward-based training

Exploratory and object play:

(a) Opportunities to climb, perch, and scratch with adjustments in height, textures, and lighting for physical limitations/mobility issues
(b) Exploratory enrichment—paper bags, cardboard boxes, and climbs
(c) Food puzzle and manipulation toys and food search play

Resources: catvets.com, indoorpet.osu.edu

References

Araujo JA, Studzinski CM, Milgram NW (2005a) Further evidence for the cholinergic hypothesis of aging and dementia from the canine model of aging. Progr Neuro-Psychopharmacol Biol Psychiatry 29(3):411–422

Araujo JA, Studzinski CM, Head E, Cotman CW, Milgram NW (2005b) Assessment of nutritional interventions for modification of age-associated cognitive decline using a canine model of human aging. Age 27(1):27–37

Araujo JA, Landsberg GM, Milgram NW, Miolo A (2008) Improvement of short-term memory performance in aged beagles by a nutraceutical supplement containing phosphatidylserine, Ginkgo biloba, vitamin E, and pyridoxine. Canad Veterinary J 49(4):379–385

Araujo JA, Greig NH, Ingram DK, Sandin J, de Rivera C, Milgram NW (2011a) Cholinesterase inhibitors improve both memory and complex learning in aged beagle dogs. J Alzheimer's Dis 26(1):143–155

Araujo JA, Nobrega JN, Raymond R, Milgram NW (2011b) Aged dogs demonstrate both increased sensitivity to scopolamine impairment and decreased muscarinic receptor density. Pharmacol Biochem Behav 98(2):203–209

Araujo JA, Faubert ML, Brooks ML, Landsberg GM, Lobprise H (2012) NOVIFIT® (NoviSAMe®) Tablets improve executive function in aged dogs and cats: implications for treatment of cognitive dysfunction syndrome. Int J Appl Res Veterinary Med 10(1):90

Bosch MN, Bayon JG, Rodríguez MJ, Pugliese M, Mahy N (2013) Rapid improvement of canine cognitive dysfunction with immunotherapy designed for Alzheimer's disease. Curr Alzheimer Res 10(5):482–493

Cai X, Campbell N, Khan B, Callahan C, Boustani M (2013) Long-term anticholinergic use and the aging brain. Alzheimer's Dementia 9(4):377–385

Campbell S, Trettien A, Kozan B (2001) A noncomparative open-label study evaluating the effect of selegiline hydrochloride in a clinical setting. Veterinary Therapeut 2(1):24–39

Carrillo MC, Ivy GO, Milgram NW, Head E, Wu P, Kitani K (1994) L-Deprenyl increases activities of superoxide dismutase (SOD) in striatum of dog brain. Life Sci 54(20): 1483–1489

Christie LA, Opii WO, Head E, Araujo JA, De Rivera C, Milgram NW, Cotman CW (2009) Short-term supplementation with acetyl-L-carnitine and lipoic acid alters plasma protein carbonyl levels but does not improve cognition in aged beagles. Exp Gerontol 44(12):752–759

Cross AJ, George J, Woodward MC, Ames D, Brodaty H, Ilomäki J, Elliott RA (2016) Potentially inappropriate medications and anticholinergic burden in older people attending memory clinics in Australia. Drugs Aging 33(1):37–44

Dakshinamurti K, Sharma SK, Geiger JD (2003) Neuroprotective actions of pyridoxine. Biochim Biophys Acta (BBA)-Proteins Proteomics 1647(1):225–229

Davis PR, Giannini G, Rudolph K et al (2016) Aβ vaccination in combination with behavioral enrichment in aged beagles: effects on cognition, Aβ and microhemorrhages. Neurobiol Aging. doi:10.1016/j.neurobiolaging.2016.09.007. Accepted 15 Sept 2016

DePorter T, Landsberg GM, Horwitz D (2015) Tools of the trade: psychopharmacology and nutrition. Feline Behav Health Welfare 245
Dodd CE, Zicker SC, Jewell DE, Fritsch DA, Lowry SR, Allen TA (2003) Can a fortified food affect the behavioral manifestations of age-related cognitive decline in dogs? Veterinary Med 98(5):396–408
Dong JY, Qin LQ, Zhang Z, Zhao Y, Wang J, Arigoni F, Zhang W (2011) Effect of oral L-arginine supplementation on blood pressure: a meta-analysis of randomized, double-blind, placebo-controlled trials. Am Heart J 162(6):959–965
Donini LM, De Felice MR, Cannella C (2007) Nutritional status determinants and cognition in the elderly. Archiv Gerontol Geriatr 44:143–153
Dowling AL, Head E (2012) Antioxidants in the canine model of human aging. Biochim Biophys Acta (BBA)-Mol Basis Dis 1822(5):685–689
Fahnestock M, Marchese M, Head E, Pop V, Michalski B, Milgram WN, Cotman CW (2012) BDNF increases with behavioral enrichment and an antioxidant diet in the aged dog. Neurobiol Aging 33(3):546–554
Geda YE, Roberts RO, Knopman DS, Christianson TJ, Pankratz VS, Ivnik RJ, Boeve BF, Tangalos EG, Petersen RC, Rocca WA (2010) Physical exercise, aging, and mild cognitive impairment: a population-based study. Archiv Neurol 67(1):80–86
Gray SL, Anderson ML, Dublin S, Hanlon JT, Hubbard R, Walker R, Yu O, Crane PK, Larson EB (2015) Cumulative use of strong anticholinergics and incident dementia: a prospective cohort study. JAMA Intern Med 175(3):401–407
Gunn-Moore D, Moffat K, Christie LA, Head E (2007) Cognitive dysfunction and the neurobiology of ageing in cats. J Small Animal Pract 48(10):546–553
Head E (2007) Combining an antioxidant-fortified diet with behavioral enrichment leads to cognitive improvement and reduced brain pathology in aging canines. Ann N Y Acad Sci 1114:398–406
Head E, Hartley J, Kameka AM, Mehta R, Ivy GO, Ruehl WW, Milgram NW (1996) The effects of L-deprenyl on spatial short term memory in young and aged dogs. Progr Neuro-Psychopharmacol Biol Psychiatry 20(3):515–530
Head E, Liu J, Hagen TM, Muggenburg BA, Milgram NW, Ames BN, Cotman CW (2002) Oxidative damage increases with age in a canine model of human brain aging. J Neurochem 82(2):375–381
Head E, Pop V, Vasilevko V, Hill M, Saing T, Sarsoza F, Nistor M, Christie LA, Milton S, Glabe C, Barrett E (2008) A two-year study with fibrillar β-amyloid (Aβ) immunization in aged canines: effects on cognitive function and brain Aβ. J Neurosci 28(14):3555–3566
Head E, Nukala VN, Fenoglio KA, Muggenburg BA, Cotman CW, Sullivan PG (2009) Effects of age, dietary, and behavioral enrichment on brain mitochondria in a canine model of human aging. Exp Neurol 220(1):171–176
Heath SE, Barabas S, Craze PG (2007) Nutritional supplementation in cases of canine cognitive dysfunction—a clinical trial. Appl Animal Behav Sci 105(4):284–296
Hill AS, Werner JA, Rogers QR, O'Neill SL, Christopher MM (2004) Lipoic acid is 10 times more toxic in cats than reported in humans, dogs or rats. J Animal Physiol Animal Nutrition 88(3-4):150–156
Houpt KA, Levine E, Landsberg GM et al. (2007) Antioxidant fortified food improves owner perceived behavior in the aging cat. In Proceedings of the ESFM feline conference, Prague
Joseph JA, Shukitt-Hale B, Willis LM (2009) Grape juice, berries, and walnuts affect brain aging and behavior. J Nutr 139(9):1813–1817
Katina S, Farbakova J, Madari A, Novak M, Zilka N (2016) Risk factors for canine cognitive dysfunction syndrome in Slovakia. Acta Veterinaria Scandinav 58(1):1
Kidd PM (2005) Neurodegeneration from mitochondrial insufficiency: nutrients, stem cells, growth factors, and prospects for brain rebuilding using integrative management. Alternat Med Rev 10(4):268
Landsberg G (2006) Therapeutic options for cognitive decline in senior pets. J Am Animal Hospital Assoc 42(6):407–413

Landsberg G, Denenberg S, Horwitz D, Mills D (2009) Behaviour problems in the senior pet. In: BSAVA manual of canine and feline behavioural medicine, 2nd edn. BSAVA, Gloucester, pp 127–135

Landsberg GM, Denenberg S, Araujo JA (2010) Cognitive dysfunction in cats: a syndrome we used to dismiss as 'old age'. J Feline Med Surg 12(11):837–848

Landsberg GM, DePorter T, Araujo JA (2011) Clinical signs and management of anxiety, sleeplessness, and cognitive dysfunction in the senior pet. Veterinary Clin N Am: Small Animal Pract 41(3):565–590

London ED, Ohata M, Takei H, French AW, Rapoport SI (1983) Regional cerebral metabolic rate for glucose in beagle dogs of different ages. Neurobiol Aging 4(2):121–126

Messonier SP (2001) Cognitive disorder (senility). The natural health bible for dogs and cats, pp 56–57

Milgram NW, Ivy GO, Head E, Murphy MP, Wu PH, Ruehl WW, Yu PH, Durden DA, Davis BA, Paterson IA, Boulton AA (1993) The effect of l-deprenyl on behavior, cognitive function, and biogenic amines in the dog. Neurochem Res 18(12):1211–1219

Milgram NW, Ivy GO, Murphy MP, Head E, Wu PH, Ruehl WW, Yu PH, Durden DA, Davis BA, Boulton AA (1995) Effects of chronic oral administration of L-deprenyl in the dog. Pharmacol Biochem Behav 51(2):421–428

Milgram NW, Head E, Zicker SC, Ikeda-Douglas C, Murphey H, Muggenberg BA, Siwak CT, Tapp PD, Lowry SR, Cotman CW (2004) Long-term treatment with antioxidants and a program of behavioral enrichment reduces age-dependent impairment in discrimination and reversal learning in beagle dogs. Exp Gerontol 39(5):753–765

Milgram NW, Landsberg G, Merrick D, Underwood MY (2015) A novel mechanism for cognitive enhancement in aged dogs with the use of a calcium-buffering protein. J Veterinary Behav: Clinl Appl Res 10(3):217–222

Mills D, Heath SE, Harrington LJ (1997) Proceedings of the first international conference on veterinary behavioural medicine. Universities Federation for Animal Welfare (UFAW), Herts, Great Britain, pp 37–41

Neus Bosch M, Pugliese M, Andrade C, Gimeno-Bayón J, Mahy N, Rodriguez MJ (2014) Amyloid-β immunotherapy reduces amyloid plaques and astroglial reaction in aged domestic dogs. Neurodegener Dis 15(1):24–37

Osella MC, Re G, Odore R, Girardi C, Badino P, Barbero R, Bergamasco L (2007) Canine cognitive dysfunction: prevalence, clinical signs treatment with a nutraceutical. Appl Animal Behav Sci 105:297–310

Overall KL (2010) Assessing brain aging in cats. DVM Newsmagazine (dvm360.com) (41) pp 6–9

Pan Y (2011) Enhancing brain functions in senior dogs: a new nutritional approach. Topics Companion Animal Med 26(1):10–16

Pan Y, Araujo JA, Burrows J et al (2013) Cognitive enhancement in middle-aged and old cats with dietary supplementation with a nutrient blend containing fish oil, B vitamins, antioxidants and arginine. Br J Nutrition 110:40–49

Pan Y, Landsberg G, Mougeot I et al (2017) Efficacy of a therapeutic diet on dogs with signs of cognitive dysfunction syndrome (CDS); a prospective double blinded placebo controlled clinical study. Proc ACVIM, National Harbor, MD

Pan Y, Larson B, Araujo JA et al (2010) Dietary supplementation with medium-chain TAG has long-lasting cognition-enhancing effects in aged dogs. Br J Nut 103:1746–1754

Parkinson FE, Rudolphi KA, Fredholm BB (1994) Propentofylline: a nucleoside transport inhibitor with neuroprotective effects in cerebral ischemia. Gen Pharmacol: Vascular Syst 25(6): 1053–1058

Penaliggon J (1997) Proceedings of the first international conference on behavioural medicine. Ed (Mills DS, Heath SE, Harrington LJ), UFAW, Herts, Great Britain 37–41

Pop V, Head E, Hill MA, Gillen D, Berchtold NC, Muggenburg BA, Milgram NW, Murphy MP, Cotman CW (2010) Synergistic effects of long-term antioxidant diet and behavioral enrichment on β-amyloid load and non-amyloidogenic processing in aged canines. J Neurosci 30(29):9831–9839

Pugliese M, Gangitano C, Ceccariglia S, Carrasco JL, Del Fà A, Rodríguez MJ, Michetti F, Mascort J, Mahy N (2007) Canine cognitive dysfunction and the cerebellum: acetylcholinesterase reduction, neuronal and glial changes. Brain Res 1139:85–94

Qian ZM, Ke Y (2014) Huperzine A: is it an effective disease-modifying drug for Alzheimer's disease? Front Aging Neurosci 6:216

Rème CA, Dramard V, Kern L, Hofmans J, Halsberghe C, Mombiela DV (2008) Effect of *S*-adenosylmethionine tablets on the reduction of age-related mental decline in dogs: a double-blinded, placebo-controlled trial. Veterinary Therapeut: Res Appl Veterinary Medicine 9(2):69–82

Ruehl WW, Bruyette DS, DePaoli A, Cotman CW, Head E, Milgram NW, Cummings BJ (1995) Canine cognitive dysfunction as a model for human age-related cognitive decline, dementia and Alzheimer's disease: clinical presentation, cognitive testing, pathology and response to 1-deprenyl therapy. Progr Brain Research 106:217–225

Ruehl WW, Neilson J, Hart B et al (1996) Therapeutic actions of l-deprenyl in dogs: a model of human brain aging. In: Goldstein D (ed) Catecholamines: bridging basic science with clinical medicine, progress in brain research. Elsevier, New York

Ruehl WW, Neilson J, Hart B, Head E, Bruyette DS, Cummings BJ (1997) Therapeutic actions of L-deprenyl in dogs: a model of human brain aging. Adv Pharmacol 42:316–319

Scarmeas N, Stern Y, Tang MX, Mayeux R, Luchsinger JA (2006) Mediterranean diet and risk for Alzheimer's disease. Ann Neurol 59(6):912–921

Selhub J, Troen A, Rosenberg IH (2010) B vitamins and the aging brain. Nutrition Rev 68:S112–S118

Siwak CT, Gruet P, Woehrlé F, Muggenburg BA, Murphey HL, Milgram NW (2000) Comparison of the effects of adrafinil, propentofylline, and nicergoline on behavior in aged dogs. Am J Veterinary Res 61(11):1410–1414

Siwak CT, Tapp PD, Milgram NW (2003) Adrafinil disrupts performance on a delayed nonmatching-to-position task in aged beagle dogs. Pharmacol Biochem Behav 76:161–168

Siwak-Tapp CT, Head E, Muggenburg BA, Milgram NW, Cotman CW (2008) Region specific neuron loss in the aged canine hippocampus is reduced by enrichment. Neurobiol Aging 29(1):39–50

Sullivan PG, Brown MR (2005) Mitochondrial aging and dysfunction in Alzheimer's disease. Progr Neuro-Psychopharmacol Biol Psychiatry 29(3):407–410

Taha AY, Henderson ST, Burnham WM (2009) Dietary enrichment with medium chain triglycerides (AC-1203) elevates polyunsaturated fatty acids in the parietal cortex of aged dogs: implications for treating age-related cognitive decline. Neurochem Res 34(9):1619–1625

Tsakiris S, Deliconstantinos G (1985) Phosphatidylserine and calmodulin effects on Ca 2+-stimulated ATPase activity of dog brain synaptosomal plasma membranes. Int J Biochem 17(10):1117–1119

Valenzuela M, Sachdev PS (2009) Harnessing brain and cognitive reserve for the prevention of dementia. Ind J Psychiatry 51(5):16–21

Valenzuela MJ, Dean SK, Sachdev P, Tuch BE, Sidhu KS (2008) Neural precursors from canine skin: a new direction for testing autologous cell replacement in the brain. Stem Cells Development 17(6):1087–1094

Veron AD, Mengoli M, Bienboire-Frasini D et al (2014) Are nasal stem cells a promising approach in geriatric veterinary medicine. In Proceedings of international congress on semiochemistry and management of animal populations, Apt France, p 63

Vivitonin®. MSD Animal Health product insert. http://www.msd-animal-health.co.nz/binaries/Vivitonin_website_label_Aug_10__tcm51-37350.pdf

Yang G, Wang Y, Tian J et al (2013) Huperzine A for Alzheimer's disease: a systematic review and meta-analysis of randomized clinical trials. PloS one 8(9):e74916

Zhang JH, Sampogna S, Morales FR, Chase MH (2005) Age-related changes in cholinergic neurons in the laterodorsal and the pedunculo-pontine tegmental nuclei of cats: a combined light and electron microscopic study. Brain Res 1052(1):47–55

Additional Resources

Enrichment guidelines.: catvets.com, icatcare.org, indoorpet.osu.edu

Landsberg G (2014) Dogs with an AARF card. Growing old with grace. In: American College of Veterinary Behaviorists et al (eds) Decoding your dog: the ultimate experts explain common dog behaviors and reveal how to prevent or change unwanted ones, Houghton Mifflin Harcourt

Landsberg GM, Denenberg S (2015) Behavior problems of the senior cat. In: Rodan I, Heath S (eds) Feline behavioral health and welfare. Elsevier, St. Louis, pp 344–356

Landsberg GM, DePorter T, Araujo JA (2011) Management of anxiety, sleeplessness and cognitive dysfunction in the senior pet. Vet Clin N Am Sm Anim Pract 565–590

Landsberg GM, Nichol J, Araujo JA (2012) Cognitive dysfunction syndrome: a disease of canine and feline brain aging. Vet Clin Small Anim 42:749–768

Landsberg GM, Hunthausen W, Ackerman LA (2013) The effect of aging on behavior in senior pets. In: Behavior problems of the dog and cat, 3rd edn. Saunders Elsevier, Edinburgh, pp 211–235

Preventive and Risk Factors of Canine Dementia

9

Aladár Mad'ari, Jana Farbakova, and Norbert Žilka

Canine cognitive dysfunction syndrome (CDS) is an age-related neurodegenerative disorder of the central nervous system. The population of senior and geriatric dogs (i.e., canines aged over 7 years) is estimated to be more than 30 million in the USA and over 15 million in Europe. As aging is the main risk factor of canine CDS, aged dogs represent a sizeable population with a great risk of developing CDS. The identification of risk and protective factors for canine dementia is important for further preventive strategies or treatment of the disease.

First of all, the human and dog share a very close bond. They share the same environment and possess similar habits (such as level of physical activity); dogs frequently even eat the same food as people. Since dogs are exposed to similar environmental influences as their owners, we might also assume that a dog's body will respond to them just as a human's would. Only limited data are available about the risk factors of CDS. Several studies showed an association between sex, size, reproductive state, nutrition, and cognitive function of aged dogs. Understanding the modifiable risk factors and thus being able to develop preventive strategies for CDS represent an important goal for veterinary medicine.

A. Mad'ari (✉) • J. Farbakova
University of Veterinary Medicine and Pharmacy, Komenskeho 73, Kosice, Slovak
e-mail: aladar_madari@yahoo.com, jana.farbakova@gmail.com

N. Žilka
Institute of Neuroimmunology, Slovak Academy of Sciences,
Dubravska 9, Bratislava, Slovak

Institute of Neuroimmunology n.o., Dvorakovo nabrezie 45, Bratislava, Slovak
e-mail: norbert.zilka@savba.sk

G. Landsberg et al. (eds.), *Canine and Feline Dementia*,
DOI 10.1007/978-3-319-53219-6_9

9.1 Canine Dementia Is an Age-Related Disorder of the Brain

With age, numerous anatomic, functional, and neurodegenerative changes are known to occur within the brain tissue, contributing to cognitive decline. Decrease in brain volume, increase in ventricular volume, demyelination, neuroaxonal degeneration, and decreased cholinergic activity are just a few hallmarks of the aging canine brain (Swanson et al. 2009). The alterations in molecular pathways contributing to age-related cognitive decline are poorly defined. Swanson et al. (2009) used canine microarrays to compare gene expression profiles of brain tissue from geriatric and young adult dogs. The report demonstrated that age had impact on gene expression, with 963 transcripts differentially expressed in geriatric dogs. As in aged rodents and humans, senior dogs displayed increased expression of genes associated with inflammation, stress response, calcium homeostasis, and, on the other hand, decreased expression of genes associated with neuropeptide signaling and synaptic transmission.

As dogs get older, their cognitive abilities gradually deteriorate, making age the key risk factor for CDS (Azkona et al. 2009; Neilson et al. 2001; Katina et al. 2016). In our studies, we have demonstrated a strong positive correlation between the age of canines and cognitive decline (Pearson correlation coefficient $r = 0.662$, $p < 0.0001$). Moreover, we found that the prevalence of CDS was similar in both small and medium/large dogs in the age range of 8–11 years (13% vs. 16%) but differed in older dogs, aged 11–13 years, with larger breeds displaying a greater prevalence of CDS (41% vs. 55%). Similarly, Azkona et al. (2009) found a gradual increase of CDS with age (9–11 years/22.7%; 12–14 years/32.7%, 15–17 years/50%). In another study, the prevalence of CDS in dogs aged 8–10 years was 3.4%, 10–12 years was 5%, 12–14 years increased to 23.3%, and over 14 years reached 41% (Salvin et al. 2010). The differences in prevalence between the studies might be caused by using various questionnaires for detection of cognitive impairment. Nevertheless, these results support the fact that increasing age exacerbates cognitive decline in dogs.

The decline of cognition is considered to be caused by age-related changes in brain metabolism and/or by specific disease processes (Christie et al. 2002; Milgram 2003; Studzinski et al. 2006; Tapp et al. 2004). The aged canine brain accumulates age-related oxidative damage of proteins and lipids (Head and Torp 2002; Kiatipattanasakul et al. 1996; Rofina et al. 2006). In parallel, endogenous antioxidant activity decreases with increasing age (Opii et al. 2008). Age-related processes may cause the reduction of the activity and levels of endogenous antioxidant enzymes, as is the case with glutamine synthetase and superoxide dismutase (Lee et al. 2014). According to the oxidative stress hypothesis, aging is linked to an accumulation of products of oxidative stress, which may be caused by a decrease in mitochondrial function and by a reduction of endogenous metabolic strategies to counteract the increase in oxidant species. The central nervous system is particularly sensitive to oxidative stress because of high metabolic rates and reduced antioxidant defenses in comparison with other tissues. This may be responsible for cognitive decline and associated neuropathology in aged dogs (Araujo et al. 2005).

9.2 Diet as Protective and Risk Factor for CDS

Impact of nutrition on dog cognition can be assessed from two points of view, for prevention of cognitive decline and as therapy of CDS. Analogously, improper diet may increase a canine's risk of suffering from CDS in older age (Katina et al. 2016). Recently, we have shown that dogs fed with uncontrolled diets had 2.8 times higher odds of developing CDS when compared with dogs fed with controlled diets. The controlled diet was characterized as high-quality commercial food purchased according to the size of the breed, age, and specific nutritional demands (e.g., low-fat diet for obese dogs) (such as Hill's, Royal Canin, Specific, etc.). An uncontrolled diet consists of kitchen waste, unspecified feed mixtures, or low-quality commercial food, which is not tailored to the size of the breed, the dog's age, or specific dietary requirements. Based on these results, we can conclude that controlled diets are a protective factor against cognitive damage, when compared to homemade or mixed diets (Katina et al. 2016).

There are several studies about a positive effect of dietary supplements and diet enriched with antioxidants or mitochondrial cofactors (Araujo et al. 2005; Fahnestock et al. 2012; Opii et al. 2008; Siwak-Tapp et al. 2008). Araujo et al. (2005) examined the effects of a diet enriched with antioxidants and mitochondrial cofactors on several measures of cognition and brain aging in a three-year longitudinal study. Aged dogs on the enriched diet demonstrated both short- and long-term cognitive benefits, as well as decreased deposition of amyloid-beta protein, and slowed down the behavioral impairment associated with CDS.

Food enriched with antioxidants and mitochondrial cofactors can ameliorate an age-related decline in mitochondrial function (Liu and Ames 2005; Head et al. 2009) and thus alleviate cognitive dysfunction (Cotman et al. 2002). Antioxidants plus mitochondrial cofactors, either alone or in combination with behavioral enrichment, attenuate lipid abnormalities in the frontal cortex of aged canines. Further analyses revealed that this type of diet can lead to a decrease in the levels of free monounsaturated fatty acid such as palmitoleic acid and nervonic acid in frontal cortex of aged dogs. In the central nervous system, fatty acids are predominantly esterified to phospholipids in the plasma membrane (Bazinet and Layé 2014). The monounsaturated/saturated fatty acid ratio, also known as "desaturation index"—an ex vivo indicator of stearoyl-CoA desaturase activity—was also reduced in the frontal cortex of dogs fed with antioxidants. Interestingly, increased palmitoleic acid levels and desaturation index were positively correlated with increased reversal learning errors and decreased cognitive performance (Snigdha et al. 2012).

9.3 Dog Size as a Risk Factor for CDS

The species *Canis familiaris* exhibits the most morphological variations of all mammalian species (Neff and Rine 2006), which coincide with variations of life span (Galis et al. 2007). Large breeds have a shorter life expectancy than small ones; this is tied to body weight rather than height (Greer et al. 2007). On the basis of differences in the life span of dogs, there is an assumption that large and heavy dogs age faster (Egenvall et al. 2005).

The oxidative stress theory states that smaller animals have a faster metabolism than large animals (e.g., mouse vs. elephant); therefore, smaller ones are more oxidatively loaded. This leads to early cell death, thus reducing life span, although this particular theory deals with interspecies differences and it is less applicable in the case of intraspecies variation. Animals with a faster growth have an increased rate of production of reactive oxygen species (ROS) in their youth, which may have an impact on life expectancy. This theory is supported by evidence from giant dog breeds, in which we can observe a greater incidence of diseases related to the increased production of ROS, such as age-related cataract (Urfer et al. 2011). Oxidative stress and ROS production are a risk factor that can also be very closely linked to cognitive aging in humans (Trollor and Valenzuela 2001) and dogs alike (Head et al. 2008).

The impact of size, breed, and life expectancy on canine cognition has been studied by different ways. For instance, some authors evaluated the effect of size (Salvin et al. 2010), while others were focused more on weight (Fast et al. 2013; Azkona et al. 2009; Neilson et al. 2001). Azkona et al. (2009) and Fast et al. (2013) divided dogs into two categories based on weight, namely, dogs up to 15 kg and dogs over 15 kg. Azkona et al. (2009) reported greater odds of showing age-related cognitive impairment in dogs weighing up to 15 kg, but they did not consider the weight like a predisposing factor. Neilson et al. (2001), Salvin et al. (2010), and Fast et al. (2013) did not describe any impact of body size on memory impairment. Similarly, our recent study revealed that size of the dog had no effect on cognitive impairment (Katina et al. 2016). In order to explore the relationship between the size and vulnerability to CDS, one needs to take the pronounced variability between dog breeds into account; often, the adult body weight of miniature breeds does not exceed 2 kg, and, vice versa, various dog breeds may easily exceed 50 kg in their adulthood. Therefore, it appears to be necessary to divide the canine population into several well-defined weight/size categories that would reflect this excessive variability.

9.4 The Influence of Environment on Canine Cognitive Status

The environment shared by humans and their animals represents highly dynamic conditions. Environmental factors that can affect cognitive decline or maintain stable cognition throughout life are essentially unique for each individual. An analysis of this shared environment could provide the key to the discovery of factors that contribute to the behavioral and functional changes in the brain of aging individuals, as well as factors which may have a protective function.

Age-related cognitive decline is affected by many external factors, some of which are less influenced by personal choice, including exposure to toxins or stressors such as noise or light pollution. The impact of environmental factors on cognitive decline is supported by evidence from studies which investigate changes in physiology associated with prolonged exposure to air pollution and stress. Inflammatory reactions of the immune system and increased oxidative stress in the mitochondria may lead to the formation of pathological changes in the brain tissue that contribute to the neurodegenerative changes in age-related cognitive decline (Block and Calderon-Garciduenas 2009). Calderon-Garciduenas et al. (2008) suggest that ultrafine particles of matter emitted from vehicle engines and emissions from aviation can contribute to neuroinflammation and oxidative stress. The authors studied wild dogs from areas with high air pollution, and they provide evidence of early occurrence of β-amyloid plaques and elevation of proinflammatory markers. Dogs exposed to high levels of pollution had diffuse β-amyloid plaques present in the brain tissue several years earlier than dogs that were living in an environment with clean air. They compared the neurophysiology of children and dogs in Mexico City and urban area with excessive air pollution. Evaluation of brain tissue by magnetic resonance imaging revealed that 56% of children and 57% of dogs from the given area present with lesions in the prefrontal subcortical white matter. Other factors may also contribute to this observed high prevalence of brain damage, including noise, heavy metals, and pesticides (Migliore and Coppede 2009).

9.5 Other Risk Factors for CDS

A high incidence of CDS combined with the fact that age and diet don't explain all the variability in incidence in aged dogs emphasizes the importance of identifying other potential individual risk factors, such as sex or reproductive status. Azkona et al. (2009) presumed that females are more likely to suffer from CDS, while other authors did not confirm this claim (Neilson et al. 2001; Yalcin et al. 2010; Salvin et al. 2011; Fast et al. 2013; Katina et al. 2016). Another putative factor involved in the development of CDS symptomatology could be the reproductive and hormonal status of dogs (castration). Two studies demonstrated a higher risk for the development of CDS in castrated individuals than in intact ones (Hart 2001; Azkona et al. 2009). On the contrary, two recent studies found no harmful effect of castration on cognitive function of older dogs (Fast et al. 2013; Katina et al. 2016). These results might be influenced by using various diagnostic approaches (questionnaires) and by a low number of animals included in the study.

Another factor with profound influence on cognitive decline is psychological stress caused by physical and emotional factors, resulting in physical, behavioral, endocrine, and immune changes. Acute and chronic stress may lead to deterioration of health and behavior. Senior dogs might be more susceptible to stress and less able to adapt to changes (Landsberg et al. 2012). Therefore, dog owners should avoid unnecessary stressful situations and possibly eliminate unnecessary stress stimuli from their environment. It is also important to give the dog the opportunity to express their normal behavior (e.g., pawing the ground, retrieving, playing with other dogs, etc.) (Manteca 2011).

9.6 Protective Factors

Protective factors for CCD can protect the brain from or can slow down the process of brain aging with cognitive decline. Most known protective factors are diet and dietary supplements. Nowadays it is possible to choose quality dog food for every dog category (puppy, adult, mature, neutered, etc.) from various brands. Some of them are specialized to old dogs, and they add nutrients with protective function from aging changes (Landsberg et al. 2012; Katina et al. 2016).

Diets rich in fruits and vegetables have been shown to improve human and animal well-being and to significantly delay the development of pathologic processes, including neurodegenerative disorders (Joseph et al. 1999; Martin et al. 2002). Foods are important sources of micronutrients, including vitamins E and C, which play crucial roles in optimal cell function. Vitamin E is an important component of biologic membranes, and vitamin C acts as a co-substrate for several enzymes. Both E and C are involved in the antioxidant defense of cells and actively contribute to the redox status of the cell (Fryer 1998; Dysken et al. 2014). Antioxidant diet can result in improvements in learning and memory and reduce the extent of pathology that accumulates in the aged brain (Cotman et al. 2002). Alpha-lipoic acid and L-carnitine reduce oxidative damage to cells as well (Joseph et al. 2000). Antioxidants also have anti-inflammatory properties. There has been an association of nonsteroidal anti-inflammatory intake and decreased incidence of dementia in humans, which suggest that inflammation is a contributor to neurocognitive decline (Cotman et al. 2002).

An environment, where people and their animals live, is a dynamic complex, such as we mentioned in a previous text, and is unique for each individual. Many studies show that the best results were achieved when antioxidant diet was combined with environmental enrichment in the form of social enrichment (contact with other dogs), physical exercise (outdoor walks in laboratory dogs or longer walks at new places in pet dogs), and cognitive enrichment (some tasks) (Head et al. 2009; Fahnestock et al. 2012; Snigdha et al. 2012).

Mitochondrial function was significantly improved in the antioxidant-fed dogs and not in environmentally enriched dogs. Interestingly, behavioral enrichment but not the antioxidant diet protected against neuron loss in the hilus of the dog hippocampus. Further brain-derived neurotrophic factor mRNA increased in aged dogs provided with the combination treatment (Fahnestock et al. 2012; Dowling and Head 2012). These facts indicate a combination of quality dog food with added protective nutrients and environmental/behavioral enrichment is the best way to protect from pathological cognitive decline of aging dogs (Tables 9.1 and 9.2).

Table 9.1 List of detected CCD risk factors

Reference/ CCD risk factors	Neilson et al. (2001)	Calderon-Garciduenas et al. (2008)	Azkona et al. (2009)	Salvin et al. (2010)	Rosado et al. (2012a)	Rosado et al. (2012b)	Fast et al. (2013)	Katina et al. (2016)
Age	+	n.a.	+	+	+	+	n.a.	+
Sex	+	n.a.	+	n.a.	n.a.	n.a.	–	–
Female in.	–	n.a.	+	n.a.	n.a.	n.a.	–	–
Male in.	–	n.a.	–	n.a.	n.a.	n.a.	–	–
Female c.	–	n.a.	+	n.a.	n.a.	n.a.	–	–
Male c.	+	n.a.	+	n.a.	n.a.	n.a.	–	–
Size	–	n.a.	–	–	n.a.	n.a.	–	–
Weight	–	n.a.	–	n.a.	n.a.	n.a.	–	–
Height	n.a.	n.a.	n.a.	–	n.a.	n.a.	n.a.	n.a.
Diet	n.a.	n.a.	n.a.	n.a.	n.a.	n.a.	n.a.	+
Air pollution	n.a.	+	n.a.	n.a.	n.a.	n.a.	n.a.	n.a.

CCD canine cognitive dysfunction; *in.* intact dogs; *c.* castrated dogs
+ Parameter was detected, and it was shown to be a CCD risk factor
– Parameter was detected, but it was not shown to be a CCD risk factor
n.a. Parameter was not assessed in the given study

Table 9.2 List of protective factors

Protective factor	Canine study	Feline study
Vitamin B_{12}	–	Pan et al. (2013)
Vitamin E	Siwak-Tapp et al. (2008)	–
	Snigdha et al. (2012)	
Vitamin C	Milgram et al. (2004)	–
	Snigdha et al. (2012)	
Mitochondrial cofactors	Cotman et al. (2002)	–
(L-carnitine and lipoic acid)	Fahnestock et al. (2012)	
MCT oil (source of ketones)	Pan et al. (2010)	–
Behavioral	Opii et al. (2008)	Pan et al. (2013)
Enrichment	Head et al. (2009)	

9.7 Summary

Cognitive decline of aged dogs represents a serious medical and social problem. The increasing number of senior dogs is accompanied by a rise in prevalence of canine CDS worldwide. Dogs are living longer than ever before, thanks to the increasing knowledge about nutritional needs and advances in veterinary medicine. However, there is a price to be paid for these additional years, and that price is an increase in geriatric disorders. The rising incidence of canine cognitive dysfunction highlights the importance of identification of putative risk factors in disease etiology.

Acknowledgment The work was funded by APVV-15-0613 and APVV-14-0872 and VEGA grant 2/0164/16.

References

Araujo JA, Studzinski CM, Head E, Cotman CW, Milgram NW (2005) Assessment of nutritional interventions for modification of age-associated cognitive decline using a canine model of human aging. Age 27(1):27–37

Azkona G, Garcia-Belenguer S, Chacon G, Rosado B, Leon M, Palacio J (2009) Prevalence and risk factors of behavioural changes associated with age-related cognitive impairment in geriatric dogs. J Small Animal Pract 50(2):87–91

Bazinet RP, Layé S (2014) Polyunsaturated fatty acids and their metabolites in brain function and disease. Nat Rev Neurosci 15(12):771–785

Block ML, Calderon-Garciduenas L (2009) Air pollution: mechanisms of neuroinflammation and CNS disease. Trends Neurosci 32:506–516

Calderon-Garciduenas L, Mora-Tiscareno A, Ontiveros E, Gomez-Garza G, Barragan-Mejia G, Broadway J, Chapman S, Valencia-Salazar G, Jewells V, Maronpot RR, Henriquez-Roldan C, Perez-Guille B, Torres-Jardon R, Herrit L, Brooks D, Osnaya-Brizuela N, Monroy ME, Gonzalez-Maciel A, Reynoso-Robles R, Villarreal-Calderon R, Solt AC, Engle RW (2008) Air pollution, cognitive deficits and brain abnormalities: A pilot study with children and dogs. Brain Cognit 68:117–127

Christie LA, Studzinski C, Estrada J, Holowachuk DZ, Leung C, Milgram NW (2002) Cognitive changes during early and late development in the dog last learned, first forgotten? Society for Neuroscience Abstract Viewer and Itinerary Planner, Abstract No. 374.378

Cotman CW, Head E, Muggenburg BA, Zicker S, Milgram NW (2002) Brain aging in the canine: a diet enriched in antioxidants reduces cognitive dysfunction. Neurobiol Aging 23(5):809–818

Dowling ALS, Head E (2012) Antioxidants in canine model of human aging. Biochim Biophys Acta 1882:685–689

Dysken MW, Sano M, Asthana S, Vertrees JE, Pallaki M, Llorente M, Love S, Schellenberg GD, McCarten JR, Malphurs J, Prieto S, Chen P, Loreck DJ, Trapp G, Bakshi RS, Mintzer JE, Heidebrink JL, Vidal-Cardona A, Arroyo LM, Cruz AR, Zachariah S, Kowall NW, Chopra MP, Craft S, Thielke S, Turvey CL, Woodman C, Monnell KA, Gordon K, Tomaska J, Segal Y, Peduzzi PN, Guarino PD (2014) Effect of vitamin E and memantine on functional decline in Alzheimer disease. The TEAM-AD VA Cooperative Randomized Trial. J Am Med Assoc 311(1):33–44

Egenvall A, Bonnett BN, Hedhammar A, Olson P (2005) Mortality in over 350,000 insured Swedish dogs from 1995-2000: II. Breed-specific age and survival patterns and relative risk for causes of death. Acta Veterinaria Scandin 46(3):121–136

Fahnestock M, Marchese M, Head E, Pop V, Michalski B, Milgram WN, Cotman CW (2012) BDNF increases with behavioral enrichment and an antioxidant diet in the aged dog. Neurobiol Aging 33(3):546–554

Fast R, Schutt T, Toft N, Moller A, Berendt M (2013) An observational study with long-term follow-up of canine cognitive dysfunction: clinical characteristics, survival, and risk factors. J Veterinary Intern Med 27(4):822–829

Fryer M (1998) Vitamin E status and neurodegenerative disease. Nutritional Neurosci 1:327–351

Galis F, Van Der Sluijs I, Van Dooren TJM, Metz JAJ, Nussbaumer M (2007) Do large dogs die young? J Exp Zool Part B-Mol Develop Evolut 308B(2):119–126

Greer KA, Canterberry SC, Murphy KE (2007) Statistical analysis regarding the effects of height and weight on life span of the domestic dog. Res Veterinary Sci 82(2):208–214
Hart BL (2001) Effect of gonadectomy on subsequent development of age-related cognitive impairment in dogs. J Am Veterinary Med Assoc 219:51–56
Head E, Torp R (2002) Insights into Aβ and presenilin from a canine model of human brain aging. Neurobiol Dis 9(1):1–10
Head E, Rofina J, Zicker S (2008) Oxidative stress, aging, and central nervous system disease in the canine model of human brain aging. Veterinary Clin N Am-Small Animal Pract 38(1):167–178
Head E, Nukala VN, Fenoglio KA, Muggenburg BA, Cotman CW, Sullivan PG (2009) Effects of age, dietary, and behavioral enrichment on brain mitochondria in a canine model of human aging. Exp Neurol 220(1):171–176
Joseph JA, Shukitt-Hale B, Denisova NA, Bielinski D, Martin A, McEwen JJ, Bickford PC (1999) Reversals of age-related declines in neuronal signal transduction, cognitive, and motor behavioral deficits with blueberry, spinach, or strawberry dietary supplementation. J Neurosci 19(18):8114–8121
Joseph JA, Denisova NA, Bielinski D, Fisher DR, Shukitt-Hale B (2000) Oxidative stress protection and vulnerability in aging: putative nutritional implications for intervention. Mechanisms Ageing Develop 116(2/3):141–153
Katina S, Farbakova J, Madari A, Novak M, Zilka N (2016) Risk factors for canine cognitive dysfunction syndrome in Slovakia. Acta Veterinaria Scandin 58:17
Kiatipattanasakul W, Nakamura S, Hossain MM, Nakayama H, Uchino T, Shumiya S, Goto N, Doi K (1996) Apoptosis in the aged dog brain. Acta Neuropathol 92(3):242–248
Landsberg GM, Nichol J, Araujo JA (2012) Cognitive dysfunction syndrome a disease of canine and feline brain aging. Veterinary Clin N Am-Small Animal Pract 42(4):749–768
Lee G-S, Jeong YW, Kim JJ, Park SW, Ko KH, Kang M, Kim YK, Jung EM, Moon C, Hyun SH, Hwang KC, Kim NH, Shin T, Jeung EB, Hwang WS (2014) A canine model of Alzheimer's disease generated by overexpressing a mutated human amyloid precursor protein. Int J Mol Med 33(4):1003–1012
Liu J, Ames BN (2005) Reducing mitochondrial decay with mitochondrial nutrients to delay and treat cognitive dysfunction, Alzheimer's disease, and Parkinson's disease. Nutritional Neurosci 8(2):67–89
Manteca X (2011) Nutrition and behavior in senior dogs. Topics Companion Animal Med 26:33–36
Martin A, Cherubini A, Andres-Lacueva C, Paniagua M, Joseph J (2002) Effects of fruits and vegetables on levels of vitamins E and C in the brain and their association with cognitive performance. J Nutrition Health Aging 6(6):392–404
Migliore L, Coppede F (2009) Environmental-induced oxidative stress in neurodegenerative disorders and aging. Mutat Res-Genetic Toxicol Environ Mutagenesis 674:73–84
Milgram NW (2003) Cognitive experience and its effect on age-dependent cognitive decline in beagle dogs. Neurochem Res 28(11):1677–1682
Milgram NW, Head E, Zicker SC, Ikeda-Douglas C, Murphey H, Muggenberg BA, Siwak CT, Tapp PD, Lowry SR, Cotman CW (2004) Long-term treatment with antioxidants and a programm of behavioral enrichment reduses age-dependent impairment in discrimination and reversal learning in beagle dogs. Exp Gerontol 39:753–765
Neff MW, Rine J (2006) A fetching model organism. Cell. 124(2):229–231
Neilson JC, Hart BL, Cliff KD, Ruehl WW (2001) Prevalence of behavioral changes associated with age-related cognitive impairment in dogs. J Am Veterinary Med Assoc 218(11):1787–1791
Opii W, Joshi G, Head E, Milgram NW, Muggenburg BA, Kleine JB, Pierce WM, Cotman CW, Butterfield DA (2008) Proteomic identification of brain proteins in the canine model of human aging following a long-term treatment with antioxidants and a program of behavioral enrichment: relevance to Alzheimer's disease. Neurobiol Aging 29(1):51–70

Pan Y, Larson B, Araujo JA, Lau W, de Rivera C, Santana R, Gore A, Milgram NW (2010) Dietary supplementation with medium-chain TAG has long-lasting cognition-enhancing effects in aged dogs. Br J Nutrition. 103(12):1746–1754

Pan YL, Araujo JA, Burrows J, de Rivera C, Gore A, Bhatnagar S, Milgram NW (2013) Cognitive enhancement in middle-aged and old cats with dietary supplementation with a nutrient blend containing fish oil, B vitamins, antioxidants and arginine. Br J Nutrition 110(1):40–49

Rofina JE, van Ederen AM, Toussaint MJM, Secreve M, van der Spek A, van der Meer I, van Eerdenburga FJCM, Gruys E (2006) Cognitive disturbances in old dogs suffering from the canine counterpart of Alzheimer's disease. Brain Res 1069(1):216–226

Rosado B, Gonzalez-Martinez A, Pesini P, Garcia-Belenguer S, Palacio J, Villegas A, Suarez ML, Santamarina G, Sarasa M (2012a) Effect of age and severity of cognitive dysfunction on spontaneous activity in pet dogs—Part 1: locomotor and exploratory behaviour. Veterinary J 194(2):189–195

Rosado B, Gonzalez-Martinez A, Pesini P, Garcia-Belenguer S, Palacio J, Villegas A, Suarez ML, Santamarina G, Sarasa M (2012b) Effect of age and severity of cognitive dysfunction on spontaneous activity in pet dogs—Part 2: social responsiveness. Veterinary J 194(2):196–201

Salvin HE, McGreevy PD, Sachdev PS, Valenzuela MJ (2010) Under diagnosis of canine cognitive dysfunction: a cross-sectional survey of older companion dogs. Veterinary J 184:277–281

Salvin HE, McGreevy PD, Sachdev PS, Valenzuela MJ (2011) The canine cognitive dysfunction rating scale (CCDR): a data-driven and ecologically relevant assessment tool. Veterinary J 188:331–336

Siwak-Tapp CT, Head E, Muggenburg BA, Milgram NW, Cotman CW (2008) Region specific neuron loss in the aged canine hippocampus is reduced by enrichment. Neurobiol Aging 29(1):39–50

Snigdha S, Astarita G, Piomelli D, Cotman CW (2012) Effect of diet and behavioral enrichment on free fatty acids in the aged canine brain. Neuroscience 202:326–333

Studzinski CM, Christie L-A, Araujo JA, Burnham WM, Head E, Cotman CW, Milgram NW (2006) Visuospatial function in the beagle dog: an early marker of cognitive decline in a model of human aging and dementia. Neurobiol Learning Memory 86(2):197–204

Swanson KS, Vester BM, Apanavicius CJ, Kirby NA, Schook LB (2009) Implications of age and diet on canine cerebral cortex transcription. Neurobiol Aging 30(8):1314–1326

Tapp D, Chu Y, Vu L, Chiou JY, Milgram NW, Nalcioglu O, Su MY (2004) Age-dependent changes in regional brain volume and cerebral blood volume in white matter of the canine brain measured using dynamic susceptibility contrast MRI. Neurobiol Aging 25:380–S381

Trollor JN, Valenzuela MJ (2001) Brain ageing in the new millennium. Austr New Zealand J Psychiatry 35(6):788–805

Urfer SR, Greer K, Wolf NS (2011) Age-related cataract in dogs: a biomarker for life span and its relation to body size. Age 33(3):451–460

Yalcin E, Kasap S, Demir G (2010) Prevalance of behavioral changes associated with age-related cognitive impairment in geriatric dogs, Bursa, Turkey. J Animal Veterinary Adv 9:1828–1832

Cognitive Dysfunction Syndrome: Information for Pet Owners

What is cognitive dysfunction?

Cognitive dysfunction is a gradual loss of brain function seen in older dogs and cats that is similar to the early stages of Alzheimer's dementia in people. There is a buildup of abnormal proteins in the brain, a loss of brain cells, an increase in toxic free radicals, and abnormal brain function. While cognitive dysfunction does not affect all pets, 50% or more of senior pets are likely to eventually develop some of the signs. Early identification and diagnosis can improve signs and possibly slow further decline.

G. Landsberg et al. (eds.), *Canine and Feline Dementia*,
DOI 10.1007/978-3-319-53219-6

Cognitive health checklist: What signs might I see in my pet?

With natural aging you may expect your pet to be less active; however if you see any of the following signs, they may indicate the onset of cognitive dysfunction.

- **Confusion and disorientation**
  - Walking into things, getting "stuck," or goes to hinge side of door
  - Staring blankly into space or looking confused, distant, or lost
  - Gets lost in the home or yard
  - Not recognizing familiar people/pets
- **Change in social contact**
  - Decreased greeting or less interest in play
  - More fearful, irritable with family members, visitors, other animals
- **Reversed sleep/wake cycle**
  - Restless/waking at night—sleeping more during the day
- **House-soiling/learning and memory**
  - Eliminating in random spots indoors—urinating in their sleeping area
  - Less able to learn tasks or remember previously learned tasks/commands
- **Change in activity level**
  - Pacing or showing repetitive behaviors
  - Decreased interest in exploration or play
- **Agitation/anxiety**
  - Fearful or anxious sounds, sights, or places that were previously of no concern
  - More fearful, anxious, or irritable with people or other pets
  - Anxious when separated from family members

What should you do?

- **Twice yearly veterinary visits for senior pets**
  - Dogs and cats age much more quickly than humans.
  - A health and "behavior" check up every 6 months can help to insure early diagnosis.
- **Prompt reporting**
  - If you see any signs on the cognitive checklist or any change in behavior, report these signs to your veterinarian.
  - Early reporting allows for early diagnosis and early diagnosis provides the best opportunity to improve signs and potentially slow progress of disease.
- **Making the diagnosis: What other conditions might cause these signs?**
  - While a change in behavior might indicate the onset of cognitive dysfunction, any change in behavior could indicate an underlying health problem. Therefore to diagnose cognitive dysfunction, other causes of the signs must be ruled out.
  - These would include disease or decline in organ function (liver, kidneys, heart), altered hormone function (thyroid, adrenal, pituitary, diabetes), a decline in the senses (sight, hearing, smell), neurological disorders, and any condition that might be causing pain (e.g., arthritis, dental).

… Is there a cure?

Unfortunately there is no simple cure for cognitive dysfunction. However, by working closely with your vet and making some simple life adjustments it is possible to maintain maximal brain health, improve your pet's quality of life, and help them to live a longer, happier life.

How you can help

- **Keep it simple**
  - Try to keep the environment the same—try not to rearrange furniture.
  - Avoid clutter and obstacles around the house and keep pathways wide and clear.
  - Keep commands and cues simple as possible and keep up the rewards for success.
- **Keep your routine**
  - Keep a predictable daily routine for your pet, particularly feeding, walking, and toileting schedule.
- **Use it or lose it**
  - Encourage your pet to use their brain more—this might reduce the onset, slow the progress, and improve the signs of cognitive dysfunction.
  - Encourage gentle exercise and play with your pet on a regular basis.
  - Consider purchasing (or making your own) puzzle toys that can be filled with food and treats to enrich your pet mentally and physically.
  - Keep up the reward-based training to help keep the brain active and healthy.
- **Know your pet**
  - Know your pet's limits when introducing new people, animals, or environments. Go slowly and make adjustments to minimize fear and help your pet to adapt. Be patient and considerate of your pet's needs. It can be a very stressful time for both of you so don't be afraid to reach out for help.

How your vet can help

- **Medication**
  - Medications can help improve cognitive function by enhancing neurotransmitters, improving blood flow, and increasing antioxidants. These might include Selegiline, Propentofylline, and Nicergoline. Note: What is available varies from country to country.
- **Diet**
  - Some prescription diets are available from your veterinarian that might improve cognitive signs and slow further decline. They act to boost antioxidant defences, reduce inflammation, improve nerve function, and provide alternative sources of energy for the aging brain.
- **Supplements**
  - A number of nutritional supplements are also available that have been shown to improve learning, memory, and signs of cognitive dysfunction and prevent or slow further damage to the brain.

The manufacturer's authorised representative in the EU is Springer Nature Customer Service Centre GmbH, Europaplatz 3, 69115 Heidelberg, Germany. If you have any concerns regarding our products, please contact ProductSafety@springernature.com

Printed and bound by CPI Group (UK) Ltd, Croydon, CR0 4YY

25/09/2026

02222863-0007